W9-BXH-290

HAZARDOUS WASTE

Recent Titles from Quorum Books

Microcomputers as Decision Aids in Law Practice
Stuart S. Nagel

Communicating Employee Responsibilities and Rights: A Modern Management Mandate
Chimezie A. B. Osigweh, Yg., editor

The Future of Hong Kong: Toward 1997 and Beyond
Hungdah Chiu, Y. C. Jao, and Yuan-li Wu, editors

Corporate Financial Planning and Management in a Deficit Economy
Louis E. V. Nevaer and Steven A. Deck

Encouraging Cooperation among Competitors: The Case of Motor Carrier Deregulation and Collective Ratemaking
William B. Tye

Political Risk Management: International Lending and Investing Under Environmental Uncertainty
Charles R. Kennedy, Jr.

Trustees, Trusteeship, and the Public Good: Issues of Accountability for Hospitals, Museums, Universities, and Libraries
James C. Baughman

Entrepreneurship and the Privatizing of Government
Calvin A. Kent, editor

The Judiciary—Selection, Compensation, Ethics, and Discipline
Marvin Comisky and Philip C. Patterson

Competitive Intelligence in the Computer Age
Carolyn M. Vella and John J. McGonagle, Jr.

How to Do Business with Russians: A Handbook and Guide for Western World Business People
Misha G. Knight

Managing Your Renovation or Move to New Offices
Robert E. Weber

Arts Administration and Management: A Guide for Arts Administrators and Their Staffs
Harvey Shore

HAZARDOUS WASTE

Confronting the Challenge

Christopher Harris, William L. Want,
and Morris A. Ward
Environmental Law Institute

Introduction by Congressman Dennis E. Eckart

Q

QUORUM BOOKS
NEW YORK • WESTPORT, CONNECTICUT • LONDON

Library of Congress Cataloging-in-Publication Data

Harris, Christopher, 1951–
Hazardous waste.

Includes index.
1. Hazardous wastes—Law and legislation—
United States. I. Want, William L. II. Ward,
Morris A. III. Title.
KF3946.H37 1987 344.73'04622 86-30571
347.3044622
ISBN 0-89930-223-8 (lib. bdg. : alk. paper)

Library of Congress Catalog Card Number: 86-30571
ISBN: 0-89930-223-8

First published in 1987 by Quorum Books

Greenwood Press, Inc.
88 Post Road West, Westport, Connecticut 06881

Printed in the United States of America

The paper used in this book complies with the Permanent Paper Standard issued by the National Information Standards Organization (Z39.48-1984).

10 9 8 7 6 5 4 3 2 1

Copyright Acknowledgments

Permission to reprint excerpts from Anne M. Burford, *Are You Tough Enough?* has been granted by McGraw-Hill Book Company.

Contents

Foreword

Congress in 1984 took some extraordinary steps in its approach to legislating in the environmental field. It was, indeed, an extraordinary time, and the 1984 Hazardous and Solid Waste Amendments reflect the tenor of that period.

The nation by then had had nearly a decade and a half of experience with strong federal pollution control laws. While much remains to be done to maintain and improve U.S. air and water quality, the experience by and large had been quite favorable. The 15-year-old Clean Air Act clearly had led to substantially improved air quality throughout much of the country. The air above Pittsburgh and a score of other major industrial cities were testimony to the progress made since the beginning of the "environmental decade" in 1970.

Improvements were demonstrable too in terms of surface water quality. The 1972 Amendments to the Federal Water Pollution Control Act (now known as the Clean Water Act) had led to marked changes in surface water quality throughout much of the United States. Salmon had returned to the Willamette River in Portland, Oregon, and also to the Connecticut River. While the statutory objective of "fishable and swimmable" waters clearly remains a goal and not a reality in many parts of the country, the progress made in improving water quality seemed unquestionable. It's an accomplishment of which the country could rightly be proud, particularly given the economic and population growth which had occurred since that first "Earth Day" in 1970.

Having addressed major air pollution and water pollution challenges in 1970 and 1972, Congress in 1976 turned its attention to the solid waste dilemma. In that year, Congress passed and President Ford signed into law the Resource Conservation and Recovery Act, RCRA. Thinking back to the major public policy challenges posed by hazardous and toxic wastes in the mid-eighties, it is striking to recall that the spotlight in the mid-seventies focused primarily on the issue of mandatory can and bottle return provisions, not on hazardous waste disposal. Nonetheless, the 1976 RCRA established comprehensive authority for federal regulation of hazardous waste.

By the late 1970s and early 1980s—in particular in the wake of the passage of the major "Superfund" statute in 1980—public, media, and congressional attention were focusing increasingly on hazardous waste rather than solid waste issues (Not necessarily, mind you, because of some grand consensus that much progress had been made on the solid waste problem!). Two important factors, one by and large "political" and the other "technical," became especially important as Congress prepared in the early 80s to reauthorize the Resource Conservation and Recovery Act.

From the political or policy standpoint, the reauthorization was falling due not only at a time of increased public awareness of hazardous waste generally, but also at a time of increasing controversy over the federal government's overall management of hazardous waste and other environmental problems. The headline-grabbing controversies that characterized the Environmental Protection Agency in late 1982 and early 1983 provided the tablet on which the RCRA Amendments were to be written. As explained in this Environmental Law Institute book, one of a series of ELI publications aimed at the broad community of pollution control and environmental professionals, those political and policy controversies played a crucial role in influencing Congress as it set about to amend RCRA.

From the technical side, one crucial factor in understanding the workings of Congress in the RCRA reauthorization involves the growing understanding in Congress of the linkage betwen RCRA and the Comprehensive Environmental Response, Compensation and Liability Act of 1980, CERCLA or Superfund. The latter law had been enacted in late 1980 to clean up the hundreds of hazardous waste sites plaguing communities nationwide. The realization that there were far more Superfund sites than initially believed—and that cleanup of each would cost much more than first thought—combined with a realization that ''loopholes'' in the 1976 RCRA law were contributing to the creation of more and more hazardous waste sites. Congress in the 1984 RCRA reauthorization was determined to shut off the valve of new potential Superfund sites.

A full year in the research, writing, and review stages, *Hazardous Waste: Confronting the Challenge* is intended to be more than just a section-by-section review of the 1984 Hazardous and Solid Waste Amendments, no matter how valuable that section of the book is on its own. Rather than just explain what the new Amendments do and how they work, ELI wanted in this book to assemble a team of authors who could tell WHY Congress did what it did, and HOW decisions were reached at various stages.

For that purpose, ELI asked Bud Ward, founder and editor of ELI's monthly policy journal, *The Environmental Forum*, to team up with two experienced attorneys whose environmental backgrounds and expertise have provided them invaluable insights into the hazardous waste public policy arena. During 1983 and 1984 Christopher Harris, now in private law practice in Washington, was Counsel to the House Subcommittee on Commerce, Transportation and Tourism, which had jurisdiction over the RCRA reauthorization bill. Bill Want's environmental legal career includes his 9 years with the Department of Justice's Land and Natural Resources Division, where he led government cases in some of the landmark pollution control litigation, and his private representation of public interest organizations and of private sector companies active in hazaradous waste management. I want to extend special thanks to Bill Want for the long hours he spent in writing the book and the depth of legal analysis he provided. Without his tireless efforts, this important project could not have come to fruition. Their skills were effectively complemented by the extraordinary support they received throughout the project from Janice O'Brien, of the Environmental Law Institute and *The Environmental Forum* staff.

The unique mixture of journalistic, policy, and legal backgrounds, I think you will agree, has led to the publication of an unusually useful, insightful, and practical book on one of the nation's most sweeping and most timely pollution control statutes.

In addition to those individuals, special recognition goes also to Claire Whitney, whose insight and invaluable comments led to many additions and improvements. Nancy Ennis showed tireless effort in cheerfully typing the original versions of many chapters. The Environmental Law Institute's Linda Johnson, Elizabeth Simon, and Scott Custin proved indispensable in the editorial production process.

As a national environmental research and publishing organization dedicated to development of more effective and more efficient environmental protection and pollution control programs, the Environmental Law Institute takes pride in publication of *Hazardous Waste: Confronting the Challenge*. I think the book will prove to be an invaluable shelf-mate to ELI's 1982 publication of *Federal Regulation of Hazardous Wastes: A Guide to RCRA*, by former EPA Deputy Administrator John Quarles, at that time a member of ELI's Board of Directors. Just as importantly, this book stands also as a testament to ELI's standing commitment to better understanding and implementation of major environmental statutes, and in that respect, it is a new and continuing manifestation of ELI's commitment to serve the professional pollution control community with timely and authoritative information it can use in the interest of better managing our natural resources.

—J. William Futrell
President
Environmental Law Institute
Washington, D.C.
September 1986

Preface

On October 5, 1984, the 98th Congress of the United States, shortly before adjourning for the upcoming presidential and congressional elections, completed action on H.R. 2867, the Hazardous and Solid Waste Amendments of 1984, and sent the bill to the White House for the President's signature. Once the bill was formally enrolled in the Congress and transmitted to the White House, President Reagan had until November 9, 1984, to sign the Amendments. The President finally signed the Amendments on November 8, just two days after his landslide re-election as President. Until then the possibility of a veto of the legislation had worried and intrigued the environmental policy community in Washington. There was widespread speculation that given his 49-state electoral college victory, the newly re-elected President could call on his renewed mandate in rejecting a sweeping set of amendments which by no stretch of the imagination could be characterized as "getting the government off the backs of the people."

Throughout its first term in office, the Reagan Administration frequently, though with little success, had run counter to "conventional wisdom" in dealing with pollution control issues. For example, only a few weeks before he was to sign the 1984 Amendments, the President had attempted, unsuccessfully, to appoint former EPA Administrator Anne Gorsuch (by this time Anne Burford) to chair the National Advisory Committee on Oceans and Atmosphere. The ensuing public and congressional uproar forced Gorsuch, who earlier had been forced to resign as EPA Administrator, to withdraw her name from consideration. Would the President surprise again and—again countering conventional wisdom—veto H.R. 2867?

It was well known that the Office of Management and Budget and other federal departments were more than a little concerned over the breadth and scope of the new 1984 Amendments. At the same time, it was generally believed that President Reagan's signing of the bill—major elements of which had been accepted in concept by the President's own new EPA leadership—could help bring to an end the extraordinarily contentious relationship on environmental issues that had developed between the Reagan Administration and the Congress. If the Administration was about to close the chapter on its political troubles with the Congress over environmental issues, would the Administration risk rekindling those fires by starting off the President's second term with a veto of a pollution control law overwhelmingly passed by both houses?

In the end, it appears that the conciliatory view prevailed. Despite ample speculation about the chances for a pocket veto, serious consideration of a veto never in fact rose to the highest levels of the White House itself. While EPA officials appeared fully prepared to do battle within the Administration to encourage the signing of the Amendments, the battle never developed. The Administration itself seems to have come to the conclusion that the enactment of the Amendments provided more benefits than disadvantages. Despite their sweeping and complicated expansion of governmental authority over the business community, the President on November 8 signed the Amendments into law while celebrating his re-election victory at his California ranch.

This book is first and foremost about the Hazardous and Solid Waste Amendments of 1984 and the rationale for their enactment. *Confronting the Challenge* offers a tour (complete with roadmap) of one of the most interesting and controversial environmental statutes since the start of the Environmental Decade in 1970.

In addition to a summary of the 1984 Amendments, the book surveys the developments in hazardous waste regulation, with particular emphasis on some of the major controversies of the early 1980s. The first chapter of Part I attempts to provide an historical perspective: the post World War II chemical revolution, the quantities of hazardous waste produced, methods of disposal, and the gradual public and congressional dissatisfaction with unrestricted land disposal. The Gorsuch-Lavelle era at EPA and the demands it created for more effective regulation are the subject of the chapter that follows. Chapter 3 reviews the common law doctrines that were used to address hazardous waste problems prior to the enactment of federal legislation. Although inadequate to the task of imposing adequate controls on pollution in an industrial age, these doctrines formed the basis of modern legislation and are still used today to fill legislative gaps.

The original effort to establish federal controls over hazardous waste is the subject of Chapter 4, which begins Part II. Enacted in 1976, the Resource Conservation and Recovery Act was a relatively straightforward statute with six major components: identification and listing of hazardous wastes; a manifest system to track them; standards for generators, transporters, and treatment, storage and disposal facilities; permitting programs for these facilities; authorization for the states to administer the program; and enforcement.

Chapter 5 deals with the 1980 Amendments to RCRA, in which Congress made adjustments to the Act, rather than adopting major new policy directions. The principal changes were to authorize EPA to regulate new facilities more stringently than it did existing ones; sanction the concept of "interim status standards"; increase the penalties for violation of the statute; suspend regulation of certain wastes; and provide various additional authorities to assist EPA in carrying out its duties.

The heart of the book, the 1984 Amendments, is the subject of Part III, Chapters 6 through 22. The vast number of new requirements, their level of detail, the innovative regulatory approaches, and the sheer number of new businesses swept into the regulatory system are startling. The Congressional Budget Office estimates that implementation of the 1984 Amendments will increase compliance costs to treat and dispose of hazardous wastes from \$4.2 billion to \$5.8 billion in 1983, to between \$8.4 billion and \$11.2 billion in 1990. By all accounts, the single most important provision concerns the stringent restrictions on land disposal. In addition to an outright ban on certain forms of land disposal (such as liquids in landfills) the 1984 Amendments mandate regulatory action by EPA to end America's reliance on land disposal by the end of the decade. To ensure that its deadlines were met, Congress enacted some "hammers." Failure by EPA to make decisions about land disposal will result in automatic bans. Congressional dissatisfaction with land disposal is also reflected in the imposition of minimum technological requirements for landfills. For instance, under the new law, all new landfills and surface impoundments are required to have double-liners and leachate collection systems, and existing surface impoundments must be retrofitted to meet these requirements within four years.

In another version of the hammer provision, all hazardous waste facilities were required to certify compliance with groundwater monitoring and financial responsibility requirements by November 8, 1985. Failure to certify compliance terminated the facility's license to operate. As a result, two-thirds of the nation's 1,575 hazardous waste landfills, pits, ponds, and lagoons were forced to close because of their inability to comply with the certification requirements.

After the tour of the hammer provisions, *Confronting the Challenge* explores the loopholes—from small quantity generators to burning and blending to the loophole that didn't really exist. Congress attempted to close all of them. And in the process, it brought as many as 155,000 more hazardous waste generators into the regulatory system and vastly expanded EPA's workload.

Congress, not content with closing loopholes in the law governing hazardous wastes, also established an entirely new subtitle to regulate underground storage of gasoline and chemical products. Congress was concerned that the approximately 100,000 underground tanks currently believed to be leaking (and the 350,000 more tanks that are likely to leak over the next five years) would contaminate ground water even faster than hazardous waste dumps.

Many, many other facets of hazardous waste regulation are covered by the 1984 RCRA Amendments including (to name just a few): new and innovative technology permits, new listing and delisting requirements, state authorization revisions, mandatory inspections, waste minimization requirements, exposure assessments, export of hazardous waste, new criminal enforcement provisions, citizen suit authority, and the National Groundwater Commission.

While these and many other components of the 1984 RCRA reauthorization were intended to fulfill Congress' basic objective of "plugging the RCRA loopholes," each provision has its own origin and legislative history. *Confronting the Challenge* tells the story of the 1984 Amendments collectively and individually and describes their unique place in the evolution of environmental law.

Introduction

On September 1, 1983, Lee M. Thomas, then EPA's Assistant Administrator for Solid Waste and Emergency Response, announced that an additional 406 hazardous waste sites would be added to the Superfund National Priority List (NPL). That action created a grand total of 546 NPL sites nationwide (a number that was subsequently increased to 850 in October 1984). This meant that approximately two-thirds of all congressional districts contained at least one Superfund site. Some, like mine, contain far more. The 11th Congressional District in Ohio, which I represent, contains a third of the 21 sites in the state.

For Members of Congress representing districts with hazardous waste sites, dealing with EPA was an exercise in bureaucratic paralysis, a phenomenon that was particularly evident while the Superfund program was being conducted under the less than competent direction of EPA Administrator Anne M. Gorsuch and Assistant Administrator Rita M. Lavelle.

Even if the Superfund program *had* been properly administered, a Member of Congress, besieged by angry constituents, legitimately concerned about contamination of their drinking water, could easily become frustrated with the tedious process of clean-up. Despite our wishes to the contrary, no one could wave a magic wand over a hazardous waste site and declare it squeaky clean.

As the chapter on the Lathrop site demonstrates, the process of cleaning up a Superfund site is a long and arduous one. While EPA has developed numerous Superfund acronyms designed to conjure up an image of decisive rationality,* the fact is that EPA, in many cases, was uncertain about the best method of clean-up. Even after all the studies at a site were completed and a remedy devised, the money to finance the clean-up had to be allocated (or obtained through litigation), the contractor had to be chosen and so forth. The anxious citizens awaited results, seldom patiently.

With the media devoting generous coverage to hazardous waste horror stories (Love Canal, LiPari Landfill, Valley of the Drums, Stringfellow Acid Pits), few politicians could afford to ignore the "ticking toxic time bombs" in their communities. The demand for immediate action was particularly intense when local citizen groups became convinced that the health of their children was endangered. "Assurances" from officials in Washington whose motives were suspect could do little to calm citizens who were understandably frightened and angry.

In this highly charged atmosphere, Members of Congress and their staff learned about the relationship between the Resource Conservation and Recovery Act (RCRA) and Superfund. RCRA was designed to assure proper management of hazardous wastes through "cradle-to-grave" regulatory con-

*For example, RAMP stands for "Remedial Action Master Plan," RI/FS means "Remedial Investigation and Feasibility Study," NCP stands for National Contingency Plan," and NDD stands for "Negotiation Decision Document."

trols. Superfund, on the other hand, was originally intended to be a remedial program which concentrated on cleaning up the nation's worst hazardous waste sites. If RCRA were working the way Congress intended, no new sites would be created in the future and Superfund would be used to address finite number of sites from past disposal practices and eliminate them as a source of danger to public health.

The trouble was that theory and reality simply didn't match. RCRA and the federal regulations promulgated to implement it were riddled with loopholes. It became evident to many of us in Congress that these loopholes were simply creating a perpetual supply of new Superfund sites.

The litany of loopholes in RCRA was recited in lengthy hearings in March 1983 before the House Subcommittee on Commerce, Transportation and Tourism, chaired by my colleague, Congressman James J. Florio of New Jersey. Among these loopholes were:

—the "domestic sewage" exemption which allowed an uncalculated (but immense) quantity of untreated hazardous waste to be disposed of into America's sewer systems;

—the "recycling for beneficial use" exemption which in Times Beach, Missouri, for instance, permitted dioxin-contaminated waste oil to be used as a dust suppressant;

—the "small generator" exemption which, as a matter of federal law, made it legal for hazardous waste generators producing up to one metric ton of hazardous waste a month to dispose of their wastes in municipal landfills;

—the "burning for energy recovery" exemption which sanctioned burning hazardous wastes as fuel in boilers used for heating apartment buildings. Because the boilers had low destruction efficiencies, the fuel's hazardous constituents were emitted into the air in large quantities;

—EPA's method for determining toxic wastes which failed to identify wastes that should be regulated under RCRA;

—the EPA permit requirements which failed to require state of the art land disposal and treatment technology;

—the "interim status" system which allowed continued disposal of hazardous waste into facilities which had no barriers (such as liners) to prevent the wastes from seeping into groundwater.

One witness who catalogued the shortcomings of EPA's regulatory program was David Lennett, a staff attorney with the Environmental Defense Fund. Toward the close of his remarks, Lennett told the Subcommittee:

> The implementation of the RCRA program has been a bitter disappointment. . . . EPA has been more adept at creating loopholes and exemptions than it has been at requiring industry to improve substantially pre-RCRA management practices.
>
> You heard this morning a whole bunch of promises, what I call "touchy-feely" regulations; "We are considering it, we are thinking about doing it." We have heard that stuff from the agen-

> cy three or four years now. We have heard it in the restricting wastes from landfills area, we have heard it in the new waste listings area. I am personally tired of hearing about it.
>
> I talk to these people all the time. Nothing changes. They just don't have the people. They don't have the enthusiasm over there, and unless you provide a legal handle for people to come in and say, "If you don't do it, we are going to get a court order to make you do it," it just does not happen.

There is no doubt that the agency was well aware of the inadequacies in the RCRA program that Lennett had complained about. EPA had task forces and numerous consultants examining virtually all of the exemptions. Yet, the most visible activity by EPA was the effort by Assistant Administrator Rita Lavelle to defeat the relatively mild RCRA reauthorization bill (H.R. 6307) passed by the House on September 8, 1982. Whether it was Lavelle's persuasive lobbying tactics or the Senate's reluctance to move on environmental legislation, the result was that no RCRA reauthorization measure was brought to the floor of the Senate during the 97th Congress. After the clock ran out on the 97th, Lavelle on December 16, 1982, testified before the House Subcommittee on Natural Resources, Agriculture Research and Environment:

> We believe that most wastes can be satisfactorily managed in the land and that it can be done with a reasonable margin of safety more cheaply in this manner. On keeping with the general philosophy stated earlier, it may be that recycling or destruction is preferable from a strictly health and environmental protection standpoint, but for many wastes, the reduction in risk achieved is probably marginal and may not be worth the cost.

Meanwhile, Lavelle and her boss, EPA Administrator Anne Gorsuch, were becoming embroiled in a cascade of controversies involving not only Superfund and RCRA, but many aspects of their management of the nation's pollution control programs. Press accounts on these controversies uncovered, among other things, the following:

- The Superfund "election track," in which Superfund cleanups were specifically timed and targeted to benefit "sympathetic" Republicans;
- The "blackballing" of scientists from advisory panels because of their alleged environmental leanings or affiliation with the Democratic Party;
- The dramatic decline in enforcement cases referred by EPA to the Department of Justice;
- The reversal of the ban on disposal of containerized liquids in landfills;
- The alleged "sweetheart deals" with defendants in Superfund cases.

While this ''reign of error'' at EPA provided considerable momentum toward enactment of a stringent RCRA reauthorization bill, there were also other factors at work. For example, the 1982 elections gave House Democrats an increase of 26 seats. In addition, in August 1983, EPA released a draft of the ''Westat study,'' an EPA contractor's findings regarding the amount of hazardous waste generated in the United States. The study showed that the estimated quantity was approximately 150 million metric tons annually—nearly four times the original estimate.* Thus the stage was set for the reauthorization of a more stringent Resource Conservation and Recovery Act during the 98th Congress.

Confronting the Challenge succeeds admirably in presenting the drama of the enactment of the 1984 RCRA Amendments along with a clear and detailed explanation of each of the statutory provisions—from the ''hammer'' provisions to ''knowing endangerment.'' The authors have been key figures in the environmental arena over the past decade and the book represents an ideal blend of their legal, legislative, and journalistic backgrounds. For an understanding of the uniquely important 1984 Amendments and a true ''legislative history'' of the forces in Congress and the country that produced it, this book is indispensable.

—Dennis E. Eckart
Member of Congress
Washington, D.C.
September 1986

*Subsequent recalculations by Westat and EPA show that the ''true quantity'' is closer to 264 million metric tons.

Part I

Overview

Chapter 1

The Historical Perspective

The chemical industry in the United States traces its ancestry to America's earliest wave of immigrants. In 1635, fifteen years after the Pilgrims landed at Plymouth Rock, John Winthrop, the colonial Governor of Massachusetts, started manufacturing small amounts of saltpeter and alum.[1] Three centuries later, in 1938, America's chemical industry was producing more than 170 million pounds of synthetic-organic chemicals.[2] The massive industrialization accompanying and following World War II is frequently characterized as the start of "the chemical revolution." It was a technological watershed that transformed America's way of life.

It has been estimated that there are approximately two-million recognized chemical compounds in existence, with nearly 250,000 new compounds produced each year.[3] In 1976, less than four decades after the chemical boom had begun, the House Committee on Interstate Commerce, in its report on the bill that became the Toxic Substances Control Act, described the magnitude and impact of the chemical revolution:

> Chemicals have become a pervasive and enduring part of our environment. They are used in our manufacturing processes, and they are essential components for consumer and industrial goods. Production and use of chemicals have surged in the recent past . . . in the past ten years, the production of synthetic organic chemicals has expanded by 233 percent. In 1973, production of the top 50 chemicals alone totaled 410 billion pounds.[4]

Although total chemical production in the United States cannot be quantified with complete precision, it is probable that the amount presently exceeds two trillion pounds each year.[5]

There can be little doubt that the successes of the chemical revolution have vastly enhanced this country's standard of living. Despite persistent concentrations of poverty, the magnitude of America's economic growth since World War II is unequalled in human history. But the success has not been without its costs.

The American chemical industry, more interested in calculating the growth rate of its many products, never adequately determined the quantity of hazardous waste it was producing. The first attempt to develop an estimate did not come until 1974, when the Battelle Memorial Institute, in a report commissioned by EPA, calculated that American industry produced 8.9 million metric tons of hazardous waste.[6] That estimate was never considered credible, and five years later another EPA study, conducted by Arthur D. Little, Inc., was completed. It calculated that industry generated 33.8 million metric tons of hazardous waste each year.[7] In 1980, yet another EPA study, this time by Putnam, Hayes & Bartlett, Inc., determined that the figure was closer to 41 million metric tons.[8]

The absence of reliable information was succinctly pointed out in the *State of the Environment 1982*, a report by the highly respected Conservation Foundation:

> There are almost no data on trends in the amount of hazardous waste that has been generated or disposed of, or on the health and environmental problems that these wastes may have caused. Relatively little is known even about the current statute of hazardous waste management. Most of the available information is based on special studies undertaken at different times by different people for different purposes. There is little consistency in definitions or methodology. The conclusions of most of the studies either are based on surveys of a limited number of facilities handling hazardous waste or are crude national estimates based on such surveys in combination with professional judgments.[9]

Apparently equally frustrated with the lack of credible information, EPA commissioned another study, this time by Westat, Inc. The study, conducted in the fall of 1982 and the spring of 1983, examined 1981 data from hazardous waste generators regulated under RCRA. The original survey results, which were released in August 1983 (just as the House began debate on H.R. 2867) indicated that approximately 150 million metric tons of hazardous waste were produced each year. In April 1984, after Westat had re-examined its data, and three months before the Senate began debate on S. 757, it became apparent that even the 150-million metric ton estimate was off the mark. The revised estimate, according to Westat, was closer to 264 million metric tons.[10] Even that figure did not provide the full picture since it did not include hazardous wastes *not regulated under RCRA*, such as wastes disposed of into sewers, small quantity generator wastes, and wastes burned for energy recovery.

Although incomplete, the 264 million metric tons figure prompted a sober realization of the enormity of the hazardous waste problem. In its March 1985 cover story, *National Geographic* described the quantity of a year's worth of hazardous waste generation in terms that anyone could envision. Two hundred and sixty-four million metric tons, it said, would be enough to "fill the New Orleans Superdome almost 1,500 times over."[11] And the effect of generating all this waste for many decades? "Since 1950," according to *National Geographic*, "we've disposed of possibly six billion tons in or on the land, steadily increasing our potential exposure to chemicals that can cause cancer, birth defects, miscarriages, nervous disorders, blood diseases, and damage to liver, kidneys or genes."

Fears about improper waste disposal were not new. A quarter century after the end of World War II, it became clear to a growing number of citizens that the price to be paid for the benefits of the chemical revolution was high. The cost being exacted came in the form of unprecedented environmental degradation and a dramatically greater risk to public health. Congress responded to these concerns by enacting a series of pollution control and environmental protection laws designed to severely restrict the excesses of the chemical revolution.

Table 1

Estimated Generation Of Industrial Hazardous Waste in 1983 By State (In thousands of metric tons)

State	Quantity	Percent of National Generation	State	Quantity	Percent of National Generation
Alabama	6,547	2.5	Montana	662	0.2
Alaska	52	a	Nebraska	739	0.3
Arizona	642	0.2	Nevada	379	0.1
Arkansas	3,729	1.4	New Hampshire	431	0.2
California	17,284	6.5	New Jersey	12,948	4.9
Colorado	1,902	0.7	New Mexico	619	0.2
Connecticut	4,238	1.6	New York	9,876	3.7
Delaware	894	0.3	North Carolina	3,954	1.5
Florida	2,981	1.1	North Dakota	269	0.1
Georgia	3,338	1.3	Ohio	19,692	7.4
Hawaii	202	0.1	Oklahoma	2,673	1.0
Idaho	1,160	0.4	Oregon	969	0.4
Illinois	14,810	5.6	Pennsylvania	18,260	6.9
Indiana	10,189	3.8	Rhode Island	1,745	0.7
Iowa	1,774	0.7	South Carolina	3,669	1.4
Kansas	2,564	1.0	South Dakota	159	0.1
Kentucky	4,647	1.7	Tennessee	12,159	4.6
Louisiana	13,801	5.2	Texas	34,866	13.1
Maine	337	0.1	Utah	1,139	0.4
Maryland	2,989	1.1	Virginia	4,038	1.5
Massachusetts	4,536	1.7	Vermont	226	0.1
Michigan	12,399	4.7	Washington	5,523	2.1
Minnesota	2,212	0.8	Wisconsin	3,297	1.2
Missouri	6,046	2.3	West Virginia	5,642	2.1
Mississippi	1,816	0.7	Wyoming	572	0.2
			Total	265,595	100.0

SOURCE: Congressional Budget Office, *Hazardous Waste Management: Recent Changes and Policy Alternatives* (May 1985). Congressional Budget Office projections for 1983 based on 1981 state employment shares found in Bureau of Census, U.S. Department of Commerce, *County Business Patterns 1981* (1981).

a. Less than one-tenth of one percent.

Congressional action during the 1970s (an era that came to be called the "environmental decade") included several statutes intended to narrow the range of acceptable disposal options. The Federal Water Pollution Control Act (also known as the Clean Water Act) established the goal of ending industrial pollution of America's rivers, streams and lakes by 1985. The Clean Air Act of 1970, and its reauthorization seven years later mandated a comprehensive effort to sharply reduce the quantity of toxic pollutants dispersed into the air. The Marine Protection, Resources and Sanctuaries Act of 1972 curtailed ocean disposal of solid and hazardous waste. The Toxic Substances Control Act of 1976, whose principal mission is to prevent harmful chemicals from coming into widespread commercial use, established tight controls on the disposal of polychlorinated biphenyls (PCBs). Other statutes, such as the 1974 Safe Drinking Water Act, directed EPA to regulate the underground injection of hazardous waste and to determine acceptable levels of chemical contamination of public drinking water supplies. The Federal Insecticide, Fungicide and Rodenticide Act of 1972 was aimed at regulating poisons whose uncontrolled dispersal into the environment threatened to fulfill the prophesy of Rachel Carson's *Silent Spring.*

Yet as the requirements of these and other pollution control statutes were gradually implemented, the industrial sector's pursuit of the most economical method of waste disposal led it to rely even more heavily on land disposal. Land disposal, after all, was the least expensive short-run option for handling wastes. Its potential long-term risks to the environment—and in particular to groundwater supplies—went unrecognized for many years. The "out-of-sight/out-of-mind" philosophy prevailed.

The Public and Congressional Concern

Those issues, such as groundwater contamination, which at the time of the 1976 congressional consideration of the Resource Conservation and Recovery Act had received little attention, by the early eighties had become driving forces in Congress' deliberations on how this country should go about managing its who-knows-how-many tons of hazardous wastes. It was both what Congress *knew* or suspected about hazardous wastes and what it *did not know* and feared that heightened its concerns about the management of the federal government's hazardous waste programs.

Further fanning the flames of public and congressional anxiety was an increasing understanding of the close connection between the hazardous waste management program, the Resource Conservation and Recovery Act, and the hazardous waste site clean-up program which since 1980 has become the most visible component in the federal government's pollution control arsenal. The Comprehensive Environmental Response, Compensation and Liability Act (CERCLA or "Superfund") had been passed in December 1980 with the intent of providing \$1.6 billion to correct pollution problems at hundreds of toxic waste sites peppering the nation's landscape. But within a year, propelled by substantial media coverage and fueled by widespread accusations of mismanagement, the waste sites clean-up activity in the early 1980s had become a political *cause celebre* for many ambitious politicians.

In addition to the enormous potential corporate liabilities, likelihood of an ongoing program lasting decades, charges of political shenanigans, and the spectre of frightened neighborhoods anxious about potential and suspected health risks associated with proximity to a dump site, politicians came also to appreciate another element in the RCRA-Superfund nexus: that even as existing dumpsites were being cleaned up at costs of millions and millions of dollars, new sites could continually feed into the pipeline unless hazardous waste management practices went hand-in-hand with clean-up programs. That is, put simply, a "critical mass" in Congress came to understand that unless the traditional waste disposal practices were abandoned, an endless supply of new Superfund sites would continue to be created. In other words, without a massive restructuring of America's waste management practices, the country would have to cope with tremendous and unpredictable clean-up costs, more and more resource-intensive and costly legal disputes, and more and more "bad press."

At the same time public and congressional awareness were crossing those thresholds of understanding, it became clear that the amounts of hazardous wastes generated annually in the U.S. were multiples of earlier rough estimates. A landmark April 1984 report to EPA's Office of Solid Waste, prepared by Westat, Inc., made clear that even the expanded estimate did not include all types or sources of hazardous waste:

> The survey's most important finding is that an estimated 71 billion gallons (264 million metric tonnes) of hazardous wastes were generated during 1981, more than six times previous estimates of annual hazardous waste generation. . . . Large portions of this quantity are mixtures of hazardous and nonhazardous wastes (e.g., hazardous wastes mixed with industrial process waters), although the survey data do not estimate the actual amounts of such mixtures, nor the concentrations of specific constituents included in such mixtures. The mixtures included by the survey are, however, hazardous wastes as defined under RCRA, and are therefore properly included in the estimate of hazardous waste generation. It is important to note, however, that this survey was not designed to estimate the quantity of *all* hazardous wastes generated during 1981. The 71 billion gallon estimate includes only those hazardous wastes generated in 1981 that were to be managed in treatment, storage, and disposal processes regulated under RCRA. Additional quantities of hazardous wastes were also generated during 1981 that were treated, stored, or disposed of in processes exempt from regulation under RCRA (e.g., hazardous wastes treated exclusively in wastewater treatment tanks covered under NPDES [National Pollutant Discharge Elimina-

Table 2

Estimated National Generation Of Industrial Hazardous Waste In 1983
Ranked By Waste Quantity
(In thousands of metric tons)

Waste Type	Estimated Range		Mean	Percent
	Lower	Upper	Quantity	of Total
Nonmetallic Inorganic Liquids	68,102	96,420	82,261	31
Nonmetallic Inorganic Sludge	23,285	32,837	28,061	11
Nonmetallic Inorganic Dusts	19,455	22,784	21,120	8
Metal-Containing Liquids	14,125	25,394	19,760	7
Miscellaneous Wastes	14,438	16,393	15,415	6
Metal-Containing Sludge	13,246	15,748	14,497	6
Waste Oils	9,835	18,664	14,249	5
Nonhalogenated Solvents	11,325	12,935	12,130	5
Halogenated Organic Solids	9,321	10,246	9,784	4
Metallic Dusts and Shavings	6,729	8,738	7,733	3
Cyanide and Metal Liquids	4,247	10,520	7,383	3
Contaminated Clay, Soil, and Sand	5,092	5,830	5,461	2
Nonhalogenated Organic Solids	4,078	5,078	4,578	2
Dye and Paint Sludge	4,035	4,438	4,236	2
Resins, Latex, and Monomer	3,451	4,585	4,018	2
Oily Sludge	2,965	4,502	3,734	1
Halogenated Solvents	2,774	4,185	3,479	1
Other Organic Liquids	2,866	4,003	3,435	1
Nonhalogenated Organic Sludge	2,179	2,305	2,242	1
Explosives	508	933	720	a
Halogenated Organic Sludge	583	848	715	a
Cyanide and Metal Sludge	537	577	557	a
Pesticides, Herbicides	19	33	26	a
Polychlorinated Biphenols	1	1	1	a
Total	223,196	307,997	265,595	

SOURCE: Congressional Budget Office, *Hazardous Waste Management: Recent Changes and Policy Alternatives* (May 1985).

a. Less than one percent.

Table 3

Waste Quantities Managed in 1983
Ranked By Major Technology
(In millions of metric tons)

Technology	Description	Quantity Managed	Percent of Total
Injection Well	Injection of liquid wastes into wells or salt caverns	66.8	25
Sewer and Direct Discharge	Discharge of treated and untreated liquids to municipal sewage treatment plants, rivers, and streams	58.9	22
Surface Impoundment	Placement of liquid wastes or sludges in pits, ponds, or lagoons	49.5	19
Hazardous Waste Landfill	Placement of liquid or solid wastes into lined disposal cells that are covered by soils	34.2	13
Sanitary Landfill	Placement of wastes in unlined dump sites, which normally receive only inert, nonhazardous materials	26.7	10
Distillation	Recovery of solvent liquids from other waste contaminants through fractional distillation	10.9	4
Industrial Boilers	Burning of wastes in industrial and commercial boilers as a fuel supplement	9.5	4
Oxidation	Chemical treatment of reactive wastes	3.0	1
Land Treatment	Biodegradation of liquid wastes or sludges in soils	2.9	1
Incineration	Burning of wastes in advanced technology incinerators meeting stringent environment standards	2.7	1
Ion Exchange	Recovery of metals in solution through membrane separative techniques	0.5	a

SOURCE: Congressional Budget Office, *Hazardous Waste Management: Recent Changes and Policy Alternatives* (May 1985).

a. Less than 1 percent.

> tion System permits under the Clean Water Act], which are excluded from regulation under RCRA). Furthermore, additional quantities of wastes were generated in 1981 that have been specifically excluded by statute and/or regulation from classification as hazardous waste, even though they may exhibit characteristics of hazardous wastes [e.g., wastes generated in conjunction with ore and minerals extraction and beneficiation; wastes legitimately disposed of through sewers to Publicly Owned Treatment Works (POTWs)].[12]

By October 1984 Congress had reached a consensus that tinkering with RCRA would not accomplish a massive restructuring of the country's waste disposal practices. According to the authors of the legislation, the threat was real and could no longer be ignored. "The continued use of some methods of land disposal of some hazardous wastes presents an unwarranted and unnecessary risk to human health and the environment," the Senate bill's sponsor, John Chafee (R-Rhode Island), told the Senate on October 5, 1984, in explaining the House-Senate conference committee report on H.R. 2867. "Particularly troublesome are landfills and surface impoundments of highly toxic, mobile, or persistent wastes and wastes that have the potential to bioaccumulate."

Chafee characterized the problems associated with land disposal of hazardous wastes as "a classic example of the inability of the free marketplace to provide for the public good. Land disposal is extremely cheap when compared with the available alternatives such as incineration or chemical-physical treatment. Therefore, we should not be surprised to find that land disposal and treatment in land disposal facilities such as surface impoundments are being utilized much more frequently than the newer, high-tech options."[13]

Concerned that land disposal poses "far greater risks to society than are necessary," Chafee told the Senate that technology exists to more safely dispose of hazardous wastes. "What we do not have, and will not have as long as cheap land disposal options are available, is a viable market to support the development and expansion of new, safer treatment and disposal technologies." He said H.R. 2867 would lead to creation of "a new market and increased demand for new, safer treatment and disposal technologies."

Across Capitol Hill in the House of Representatives, the same concerns were being voiced by Republicans and Democrats alike. "For too many years in this country, we have permitted the practice of dumping hazardous wastes in the land to go virtually unchecked," Republican Norman Lent told the House in supporting the House-Senate conference report on H.R. 2867. "Even now that EPA is requiring landfills to be lined, I do not feel confident that these liners will remain secure in the long term. Therefore, I believe it is appropriate for the Congress to intervene at this time and to establish a new policy which calls for a review of known hazardous wastes and a determination whether these wastes are appropriate for land disposal," said Lent, ranking

Table 4

Comparison of Expenditures By Major Industry Groups For Hazardous Waste Management With Sales and Value Added in 1983 (In millions of dollars and percents)

Major Industry	Expenditures	Sales	Percent of Sales	Value Added	Percent of Value Added
Wood Preserving	56	1,350	4.1	432	13.0
Fabricated Metal Products	899	38,640	2.3	19,282	4.7
Chemicals and Allied Products	1,544	84,507	1.8	48,058	3.2
Rubber and Plastic Products	798	56,691	1.4	28,270	2.8
Primary Metals	1,243	123,122	1.0	44,770	2.8
Miscellaneous Manufacturing	267	29,701	0.9	15,384	1.7
Nonelectrical Machinery	254	83,839	0.3	27,563	0.5
Electrical and Electronic Machinery	156	46,967	0.3	46,507	0.5
Petroleum and Coal Products	136	175,242	0.1	29,327	0.5
Transportation Equipment	191	201,946	0.1	68,904	0.3
Motor Freight Transportation	299	a	a	b	b
Drum Reconditioners	6	a	a	b	b
Weighted Average			1.3		4.3

SOURCE: Congressional Budget Office, *Hazardous Waste Management: Recent Changes and Policy Alternatives* (May 1985).

a. Sales data not available.

b. Value added data not available.

Table 5

Estimated National Generation of Industrial Hazardous Wastes Ranked By Major Industry Group[a]
(In thousands of metric tons)

Major Industry	Estimated Quantity in 1983	Percent of Total
Chemicals and Allied Products	127,245	47.9
Primary Metals	47,704	18.0
Petroleum and Coal Products	31,358	11.8
Fabricated Metal Products	25,364	9.6
Rubber and Plastic Products	14,600	5.5
Miscellaneous Manufacturing	5,614	2.1
Nonelectrical Machinery	4,859	1.8
Transportation Equipment	2,977	1.1
Motor Freight Transportation	2,160	0.8
Electrical and Electronic Machinery	1,929	0.7
Wood Preserving	1,739	0.7
Drum Reconditioners	45	b
Total	265,595	100.0

SOURCE: Congressional Budget Office, *Hazardous Waste Management: Recent Changes and Policy Alternatives* (May 1985).

a. See CBO Report, *Empirical Analysis*, Table 1, for the master list of specific industry types that are aggregated into the major industry groups presented in this table.

b. Less than one-tenth of one percent.

Republican member on the House Energy and Commerce Committee's Subcommittee on Commerce, Transportation and Tourism.[14]

To New Jersey Democrat James J. Florio, chairman of the Commerce, Transportation and Tourism Subcommittee, passage of the bill, "if conscientiously implemented, will force a massive change in this country's disposal practices: That is, away from land disposal and toward appropriate recycling, waste reduction, and treatment."[15]

Congressman Dennis Eckart (D-Ohio), a key player in forging the compromises agreed to by the Conference Committee, summarized the Nation's new land disposal policy: "Congress will be sending a clear and unambiguous message to the regulated community and the Environmental Protection Agency: reliance on land disposal of hazardous waste has resulted in an unacceptable risk to human health and the environment. Consequently, the Congress intends that through the rigorous implementation of the objectives of this act, land disposal will be eliminated for many wastes and minimized for all others, and that advanced treatment, recycling, incineration, and other hazardous waste control technologies should quickly replace land disposal. In other words, land disposal should be used only as a last resort and only under conditions which are fully protective of human health and the environment."

Lathrop, California: A Case Study . . . The Kind of Land Disposal Congress Wanted to Stop

There are plenty of "horror stories," some well-known, involving the disposal of hazardous waste. These horror stories initiatives prompted congressional action on hazardous waste amendments at a time when other pollution control legislation was stalled on Capital Hill.

Some of the hazardous waste "horror stories," through extensive print and network news coverage, have become deeply imbedded in the public consciousness: For instance, Love Canal in New York—with its residents forced to abandon their homes through a government salvage sale rather than face further health risks from exposure to highly toxic chemicals. With a political impact which was felt across the Nation, Love Canal has come to symbolize the devastating effects of toxic wastes on families and neighborhoods.

But Love Canal was by no means alone. In Times Beach, Missouri, casual spraying of tens of thousands of gallons of contaminated used oil led to another hazardous waste horror story. The contaminated oil, which was later shown to contain extremely high dioxin concentrations, was used to suppress dust on several miles of Times Beach's unpaved roads. The resulting public health emergency once again led to the abandonment of an entire residential area and another hazardous waste site "fire sale."

One recalls a horror story from an earlier era—the June 22, 1969, nightmare of an Ohio River set ablaze as a result of its heavy pollution burden. A graphic four-color image in *Time* magazine of the polluted Cuyahoga River on fire captured the imagination and fears of a public becoming increasingly concerned about their environment. *Time* commented, "Chocolate-brown, oily, bubbling with subsurface gases, it oozes rather than flows."[16]

To many, Times Beach, Love Canal and other sites such as the "Valley of the Drums" became hazardous wastes' counterparts to the blazing Cuyohoga. They became vivid symbols of "quick and dirty" disposal practices gone awry. Yet for Members of Congress considering the reauthorization of the Resource Conservation and Recovery Act, there was more to worry about than "just" the highly-publicized horror stories. By the fall of 1983, approximately two-thirds of the House of Representatives' 435 congressional districts had at least one site on the Superfund National Priority List. Particularly for House members facing re-election every two years, a "back-home" waste site presented constant political problems. An analysis of one of those sites—at Lathrop, California—offers some insight into what Congress had on its "collective mind" when it passed the Hazardous and Solid Waste Amendments of 1984.

Nestled in California's Central Valley in San Joaquin County, about 10 miles south of Stockton and 55 miles south of Sacramento, the small unincorporated community of Lathrop was home to some 5,000 people. Many of them worked for one of the several major industries in Lathrop, others in the Central Valley's farmlands, considered among the most productive in the world.[17]

The land in the Lathrop area slopes gently toward the San Francisco Delta, falling about two feet per mile. The soil in the area is highly permeable, allowing liquids to migrate easily between surface and ground waters. Bordered on the east by the Sierra-Nevada mountain range and on the west by the Diablo range, Lathrop sits 13 to 25 feet above an aquifer. That aquifer is the Lathrop area's primary source of residential, industrial, and agricultural water. Because the annual evaporation rate in the Central Valley is about five times higher than the annual rainfall, the area makes extensive and regular use of agricultural irrigation water.

The year was 1953. The Best Fertilizer Company opened a plant in Lathrop, one half mile from a well that serves most of the town's 800 residential and business customers. The company began to use unlined surface impoundments for storing evaporating rinse water from its fertilizer and pesticide operations. Surface impoundments were also used for the systematic disposal of pesticide and fertilizer process wastes. In one unlined surface impoundment which become known as "the boneyard," Best Fertilizer (acquired by Occidental Petroleum Corporation in 1964) had disposed not only of process wastes but also of as many as 200 different off-specification pesticides that had been formulated at the plant since 1957.

Perhaps the most dangerous waste at Lathrop was the nematocide DBCP, a chemical known for its persistence in the environment, its mutagenicity, its carcinogenicity and its ability to decompose fatty tissues. Also disposed of at various locations on the 130-acre plant site were burned solid wastes, spent catalysts such as vanadium, nickel, copper, and zinc, off-specification fertilizers, spent activated carbon, and construction debris.

At hearings in 1979 on hazardous waste disposal before the House Oversight and Investigations Subcommittee of the Committee on Interstate and Foreign Commerce (now named the Energy and Commerce Committee), Robert Edson, a Lathrop plant engineer estimated that more than five tons of pesticide-related wastes had been disposed of annually into the ground at Lathrop. Also dumped were radioactive wastes from the manufacture of phosphoric acid. Edson also testified that four years earlier, in an April 29, 1975, memorandum to company officials, he had warned that the California Water Quality Control Board was unaware of the plant's pollution problems. Among other things, the memorandum stated that "all of our gypsum water [waste from phosphoric acid manufacture], our pesticide wastes, and 1 to 3 percent of our product" (pesticides and fertilizers) were percolating into the Lathrop soil. "We have destroyed the useability of several wells in our area," Edson advised company officials. Also, in that memorandum Edson recounted a story of a neighborhood dog's ill-fated venture into a field where the Lathrop plant's wastewater was percolating. In recounting the incident of the congressional hearing, Edson testified that the dog licked itself and died, presumably from poisoning.[18]

Shortly before Edson's congressional testimony, California Water Quality Control Board inspectors sampled Lathrop gypsum ponds and found that surface sulfate levels had risen from 26 micrograms per liter in 1962 to 1,700 micrograms per liter in 1979. The state inspectors also sampled a well on property adjacent to the Lathrop site and a small dairy farmer's well just south of the site. Both wells were found to be contaminated with about 13 micrograms per liter of DBCP, and the owners were advised to stop using them the same time that the DBCP contamination was found in neighboring wells. While there is no federal drinking water standard for DBCP, California's Department of Health Services had established 1 part per billion as the "action level" at which it would advise against drinking the water. Upon receiving confirmation of the DBCP contamination from its own consultant, Occidental immediately connected the dairy farmer to an uncontaminated well at Occidental's expense.

There was more. The radioactive wastes were found to have percolated into the groundwater around the Lathrop facility. These wastes had contaminated both domestic and irrigation wells with, in some cases, 10 times the federal recommended level of radioactivity. The water in the well serving Lathrop Elementary School showed uranium radioactivity at nine times the federal recommended level. School officials immediately took steps to obtain uncontaminated water supplied by the Lathrop County Water District.

As if determined to roll all of the drama into one scene, the playwright for Lathrop threw in for good measure another element certain to capture attention. Several male Lathrop plant workers who had been exposed to DBCP found they were unable to father children. Tests of the male workers exposed to DBCP revealed that more than 13 percent could produce no sperm, and that nearly 33 percent produced few sperm.[19] These test findings were enough to trigger both state and federal regulation of DBCP, and enough to convince the Lathrop facility to stop making it.

In the spring of 1979, the Central Valley Regional Water Quality Control Board found that the Lathrop plant had contaminated groundwater by "intentionally or negligently discharg[ing] waste in violation of waste discharge requirements." The Board directed the facility to halt the discharge of pesticides and waste byproducts to ground or surface waters, and to halt disposal of wastewater into ponds or other areas contaminated with the pesticide products and byproducts. By then, of course, substantial damage had been done. It now was time to think about correcting the mistakes of the past.

Long-term cleanup plans for the Lathrop site had to be developed. In this effort, the federal government, two state agencies, and Occidental developed a plan for extracting, treating, and subsequently reinjecting groundwater that had been contaminated over the years. The plan involved hydrologists, engineers, chemists, and lawyers who were to evaluate the region's hydrogeologic conditions; analyze statutory and regulatory requirements; determine the types and quantities of chemical constituents at the site, and their long-term risk factors; and determine the size of the area to be cleaned up and the costs of conducting the appropriate remedial activities.

As required under the federal National Contingency Plan adopted by EPA under the Comprehensive Environmental Response, Compensation and Liability Act, a full range of alternatives was considered:

- Taking no remedial action whatsoever, a clearly inappropriate option.
- Excavating the contaminated soils, and transporting them away from Lathrop for disposal and/or treatment at a licensed facility. This alternative raises the well-known "How Clean is Clean?" Superfund riddle: Given this alternative, precisely how much of the contaminated soil should actually be excavated? Six inches? Twelve inches? Eighteen?
- Another option, instead of excavating and removing the contaminated soils, involved containing them on-site, and utilizing various methods to prevent the wastes from spreading to adjoining areas. Underground walls and other containment barriers, impeding movement of the contamination, could be constructed. Fixation—reducing the wastes into a solid and non-volatile form—could be attempted, or the groundwater gradient could be modified to prevent movement and further contamination.

All options were considered, but excavation and off-site disposal were chosen because they offered the best hope of preventing additional groundwater contamination.[20]

Even with the off-site disposal option selected, numerous other questions remained: How could the excavated areas be sealed off so that remaining chemicals would not seep into them? Should these areas be capped with cement? With a mixture of cement and soils? With asphalt? With clay trucked in from off-site? With some combination of custom synthetic liner specifically made for that purpose? The choice was to use clay transported from off-site.

Other questions arose: What should be done with excess groundwater extracted for treatment? How would that water be treated or disposed of? Air stripping, ultraviolet oxidation and peroxide oxidation were options, as was carbon adsorption. Test results led to the selection of carbon adsorption.

Through a series of negotiations involving the Environmental Protection Agency's Denver-based National Enforcement Investigations Center, California's Water Quality Control Board and Department of Health Services, and Occidential, decisions were made as to precisely how much soil would be excavated and how extensive the groundwater treatment would be. In the absence of specific statutory or federal guidance, those kinds of decisions were based on (1) a range of pollutant standards, most of which were adopted through programs designed to protect other environmental media; (2) participants' "best professional judgments," and (3) the specific geographical and pollution factors which distinguished the Lathrop site.

As specified under the consent decree[21] agreed to by Occidental, EPA, and state authorities, the company financed the entire cleanup, including new water supply hookups for 28 residents of Lathrop whose groundwater wells were either contaminated or at risk of contamination. The state agencies were reimbursed by the company for their sampling and testing costs, and both EPA and the state were reimbursed for the costs of their investigations prior to the parties' reaching the settlement. Also, the consent decree required the company to make contributions to finance environmental research activities at state colleges and universities. The decree specified that the groundwater treatment system at Lathrop had to be maintained until the year 2001. Some 274 30-cubic-yard capacity trucks were needed to haul off contaminated Lathrop soils. The trucks hauled much of the waste soil 140 miles to a licensed Class I landfill in Coalinga, California. The less hazardous wastes were disposed of in a Stockton, California, landfill just 15 miles from the Lathrop site. In total, the Lathrop remediation process involved more than 4,600 cubic yards of contaminated soils which were excavated, transported, and disposed of at a cost of nearly one-quarter million dollars.

But the cost of remedying the Lathrop site did not stop there. Another one-and-a-half million dollars was spent to develop, design, and construct the groundwater treatment system. Another $175,000 was spent for a contractor study of the Lathrop groundwater problems. Other costs include annual operation and maintenance costs; replacement of carbon; electricity to three continuously run 7.5-horsepower pumps used in the extraction wells; and construction of alternative water supplies.

Although the Lathrop site did not receive even a fraction of the national publicity devoted to Love Canal, it was Lathrop and hundreds of similar sites that were on the minds of Members of Congress as they began to reconsider and reauthorize the 1976 Resource Conservation and Recovery Act.

To the 98th Congress, the Lathrop site was just one more example of the typical land disposal practices that had been going on for decades. These practices, in Congress' view, simply led to creation of Superfund sites which, in turn, required complicated, expensive and frequently controversial clean-up efforts.

Congress wanted it stopped.

Chapter 1 Footnotes

[1] Manufacturing Chemists Association, *The Chemical Industry Facts Book*, 1 (Fifth ed., 1962).

[2] *Id.* at 17.

[3] S. Rpt. No. 94-698, 94th Cong., 2d Sess. 3 (March 16, 1976).

[4] H.R. Rpt. No. 94-1341, 94th Cong., 2d Sess. 3 (July 14, 1976).

[5] See generally, U.S. Bureau of Census, *Current Industrial Reports*, series MA28A, MA28B, MA28C and MA30-A (1981, 1982); U.S. Bureau of the Census, 1982 *Census of Manufacturers.*

[6] Battelle Memorial Institute, *Program for the Management of Hazardous Wastes*, 1974 2v. (Distributed by the National Technical Information Service, Springfield, Virginia, as PB-233630).

[7] Arthur D. Little, Inc., *Draft Economic Impact Analysis, Subtitle C, Resource Conservation and Recovery Act of 1976*, 98 (Jan. 1979) (prepared for EPA, Office of Solid Waste).

[8] Booz-Allen & Hamilton, Inc., and Putnam, Hayes & Bartlett, Inc., *Hazardous Waste Generation and Commercial Hazardous Waste Management Capacity: An Assessment*, III-2 (Dec. 1980) (prepared for EPA, Office of Solid Waste).

[9] Conservation Foundation, *The State of the Environment 1982*, pp.145-146.

[10] Westat, Inc., *National Survey of Hazardous Waste Generators and Treatment, Storage and Disposal Facilities Regulated Under RCRA in 1981*, p.2 (April 1984) (prepared for EPA, Office of Solid Waste).

[11] *National Geographic*, March 1984, p.325.

[12] Westat, Inc., *National Survey of Hazardous Waste Generators and Treatment, Storage, and Disposal Facilities Regulated Under RCRA in 1981*, pp.2-3 (April 1984) (prepared for EPA, Office of Solid Waste).

[13] *131 Cong. Rec. S13818 (daily ed., Oct. 5, 1984).*

[14] *129 Cong. Rec. H11142 (daily ed., Oct. 3, 1984).*

[15] *129 Cong. Rec. H11141 (daily ed., Oct. 3, 1984).*

[16] "The Cities: The Price of Optimism," *Time*, pp.41-44, August 1, 1969.

[17] The descriptive material in this section on Lathrop is based primarily on two official government reports: "Six Case Studies of Compensation for Toxic Substances Pollution: Alabama, California, Michigan, Missouri, New Jersey, and Texas," a report prepared by the Environmental Law Institute under contract to the Library of Congress' Congressional Research Service, Serial No. 96-13, June 1980 (hereinafter referred to as "Six Case Studies Report"); and "Case Studies 1-23: Remedial Response at Hazardous Waste Sites," U.S. EPA, EPA-540/2-84-002b, March 1984 (hereinafter referred to as "Case Study 17").

[18] Hearings before the Subcommittee on Oversight and Investigations of the House Committee on Interstate and Foreign Commerce, 96th Cong., 1st Sess., 693-694 (1979).

[19] "Six Case Studies Report," 151.

[20] "Case Study 17," 17-18.

[21] *U.S. v. Occidental Chemical Corporation*, C.A. No. S-79-989 (E.D. Cal.).

Chapter 2

The Gorsuch Era at EPA and Its Influence

A lot goes into the making of a law in Washington. Beyond the specific technical issues at stake in environmental statutes, there is a virtually unlimited number of other issues at play, some obvious and some much more subtle.

In the case of the Hazardous and Solid Waste Amendments of 1984, Congress's deliberations were shaped by far more than "just" the mounting evidence of groundwater contamination problems facing the country. It was more than the growing number of hazardous waste sites nationwide and the alarming costs of Superfund cleanups.

It is no secret that Congress, in reauthorizing RCRA, was driven by a deep-seated distrust of the Executive Branch's handling of its environmental responsibilities. Congressional dissatisfaction with the Executive Branch is by no means new, and it certainly is not limited to environmental issues. But the seething antagonism that enveloped Capitol Hill's relations with President Reagan's environmental policy team was unprecedented.

The real and perceived shortcomings of the Reagan Administration's early handling of environmental programs provided the background for reauthorization of the Resource Conservation and Recovery Act. Congress' impressions of the Administration's environmental initiatives began to develop early in the President's first term. By the time Anne McGill Gorsuch, on February 21, 1981, was named as President Reagan's choice to be Administrator of the Environmental Protection Agency, the controversies were already well under way. Her nomination did little to calm them.

Anne Gorsuch, born in Casper, Wyoming, in 1942, was to become the fourth Administrator of the Environmental Protection Agency. Her approval by the Senate Environment and Public Works Committee came on a 16-0 vote, and the Senate followed by confirming her on a voice vote. Gorsuch was sworn into office on May 20, 1981, four months after President Reagan took office following his landslide election.

Then 39, Gorsuch had been a lawyer with the Mountain Bell Corporate Law Department in Denver. Her resume indicates that her work with Mountain Bell primarily involved land acquisitions, leases, easements, and right-of-way agreements.[1] Gorsuch from 1968 to 1971 had served as Assistant District Attorney for Jefferson County, Colorado, and from 1971 to 1973 as Deputy District Attorney for Denver. In the latter position she was responsible for all non-support claims in the Denver metropolitan area, supervising a secretarial staff of six, one paralegal, and two investigators.[2] In both the Jefferson County and Denver legal positions, Gorsuch shared work responsibilities with a friend and attorney, who said: "To my knowledge we were the first professionals—certainly the first lawyers—to implement the concept of job-sharing in Colorado."[3]

Raised in Denver after her family relocated from Wyoming shortly after she was born, Gorsuch is the daughter of a prominent and politically conservative physician. By the time she graduated in 1959 from Denver's Saint Francis de Sales High School, Gorsuch had spent three summers studying Spanish literature and Mexican history and archaeology at the National University of Mexico. She earned her bachelor of arts degree in political science from the University of Colorado at Boulder after just two years in college, graduating at age 19. Three years later, in 1964, she graduated from the University of Colorado Law School and later became the youngest woman ever admitted to the Colorado Bar. She spent 1964 and 1965 as a Fulbright Scholar in Jaipur, India, where she taught English and studied the Indian judicial system and penal institutions from the standpoint of Indian modifications to English common law.

Anne Gorsuch's political career started in 1976 while she was working for Mountain Bell. In that year she was elected to the Colorado State House of Representatives, where she served for four years. Voted by her peers as the House's "Outstanding Freshman Legislator," Gorsuch in her first term served as a member of the Finance and Appropriations Committees, and as Vice-Chairman of the Judiciary Committee. Her resume states that she "successfully sponsored 21 bills," and that she was "responsible for repeal of over 70 sections of statutes which were obsolete, duplicative, or unnecessary."[4]

Re-elected in 1978, Gorsuch was named to chair the influential House State Affairs Committee after she and several House colleagues helped engineer the ouster of the Republican Speaker of the House. In his place was put the more conservative Robert Burford, who later became a top Interior Department official in the Reagan Administration and whom Gorsuch married on February 20, 1983. In her second term in the Colorado House, Gorsuch served on the Transportation and Energy Committee and chaired the Interim Committee on Hazardous Waste. She was identified as a leader of an energetic band of conservative Colorado House Republicans known in Denver as "the crazies."

Gorsuch's Colorado legislative experiences gave her some, but not extensive, experience with pollution control programs. Since Denver was not meeting the federal public health-based air quality standards for ozone and carbon monoxide under the Clean Air Act, the state was under heavy pressure from EPA headquarters to adopt an automobile emissions inspection and maintenance program. Facing the possible imposition of building construction bans and the loss of federal grants to Colorado, Gorsuch sponsored the state law authorizing the inspection and maintenance plan in the Denver metropolitan area.

In the hazardous waste area, which later was to become the most visible aspect of her record as EPA Administrator, Gorsuch's tenure as a state legislator was highlighted by her opposition to state assumption of the federal hazardous waste program. Testifying before the Senate Environment and Public Works Committee at her confirmation hearings, Gorsuch explained that "the disadvantages simply outweighed the advantages" of the state's being delegated the hazardous waste program from EPA.[5] She thought that this experience would be useful in her new position, stating: "I fully expect that a familiarity with the major and minor elements which make assumption of a federal program undesirable will be a major asset in the job of Administrator."[6] Further she said her state legislative experiences "have given me a rich insight into the substance of some of the major environmental issues of our time, and equally importantly, the role of the states and their relationship to EPA."[7]

Testifying on her plans in office, Gorsuch said, "We can and must simplify and streamline the regulatory process."[8] She said "the public is fully committed to environmental protection, while simultaneously aware of the need to improve our economy and develop affordable domestic energy resources." To Gorsuch, it was clear that "a delicate balance must emerge."[9] EPA under her guidance could "take the lead" in achieving that balance. The agency, she claimed, would be "cognizant of its opportunities and constraints, its potential and limitations, its historic mission and our changing times." Gorsuch said it would be important for EPA to seek that balance in a "nonconfrontational" way, leading "by action and encouragement," an approach she pledged would be her "guiding credo."[10]

"I am confident that President Reagan did not ask me to serve because of any advocacy position I have taken. I have not made my living fighting for or against environmental laws and regulations," Gorsuch told the Senate Committee. "I am convinced that he has asked me to serve because he believes that my education and experience have trained me to take the broad overview necessary."[11]

However, despite Gorsuch's self-confidence, substantial intellect, and forceful presence, serious questions persisted about her professional qualifications to head the federal government's largest regulatory agency. Many expert observers of political Washington and of federal environmental policy making felt that the Colorado legislative experiences with some pollution issues by no means had adequately prepared her to handle the myriad complex programs administered by EPA. Even Mrs. Gorsuch's staunchest advocates did not seek to defend her nomination on the basis of substantive familiarity with pollution control issues, but rather because of her obvious intellectual capabilities, and her commitment to the newly elected President's apparent mandate to "get the government off the peoples' back." Gorsuch unquestionably possessed those qualities.

Nonetheless, from the time she first was mentioned in newspaper "leaks" as President Reagan's nominee for EPA Administrator, Gorsuch had to contend with the issues of professional qualifications and, in particular, of "independence" from the pervasive influence of the Office of Management and Budget and from newly named Interior Secretary James Watt, a friend of Gorsuch's from Denver. In the weeks leading up to her Senate confirmation committee hearing and her confirmation by the full Senate, Mrs. Gorsuch was portrayed by her critics as an ideological clone of James Watt, who later became the first-term Reagan Administration's most controversial Cabinet member. There were repeated, but never fully-substantiated, reports that both Gorsuch and Watt had been the hand-picked selections of conservative Colorado brewer Adolph Coors.

Mrs. Gorsuch, for whatever reasons, did little to dispel the reputation. Indeed, she seemed almost to go out of her way to contribute to the impression: Prior to her confirmation, she sequestered herself just a stone's throw from Watt's office in the Interior Department headquarters. In the highly politically charged Washington community to which she was just being introduced, the image prompted congressional anxieties about her commitment and ability to maintain an appropriate "independence" within the Administration.*

Pointing to the issue of "independence" of the EPA Administrator, an issue which by then had become the major concern of Gorsuch's critics on and off the Senate committee, New Mexico Senator Pete V. Domenici expressed confidence in her ability to oversee the agency and not succumb to inappropriate pressures within the Administration. The well-respected Domenici disagreed with those characterizing Gorsuch as "someone who can be led around by the nose"[12] "From everything I've heard, she will be her own person," he said.

That assurance was not enough for Senator John Chafee, however. The Rhode Island Republican, chairman of the Senate Environment and Public Works Committee's Subcommittee on Environmental Pollution, wanted more. For Chafee, it was not sufficient for the EPA Administrator to merely be an environmental manager. He wanted the Administrator to be an active proponent of environmental improvements. "My concern is that you not merely be a compromiser or a forger of coalitions," Chafee told Gorsuch, "but that you will be an advocate as head of the Environmental Protection Agency."

While Senator Chafee had his doubts, the support for Gorsuch's nomination was enthusiastic from many individuals and organizations outside the environmental community. "Mrs. Gorsuch's experience as a state legislator in Colorado is especially timely for EPA given President Reagan's commitment to streamline the federal government's regulations and regulatory procedures," William A. Cox, Jr., president of the 80,000-member National Society of Professional Engineers, said in a letter to Senate Environment and Public Works Committee Chairman Robert T. Stafford.[13] "Ms. Gorsuch also is quite knowledgeable regarding environmental concerns," the letter continued.

Denver Mayor William H. McNichols, Jr., endorsed Gorsuch's nomination "without reservation," saying she would bring "strong leadership and direction to the EPA and provide a balanced program of environmental pro-

*The issue of EPA's being "independent" is a perennial one between the Executive and Legislative Branches, and is by no means unique to the Reagan Administration. While the agency is unquestionably expected to be part of the President's "team," it also is the largest and most pervasive of the so-called independent agencies, and its programs are by statute and by their nature heavily regulatory. While not at all unique to the Reagan Administration, however, the "independence" issue became even more intense during the early years of the Administration than it had been previously, or, for that matter, since then.

tection to all sections of this country." [14] The Associated General Contractors of America wrote to Senator Jennings Randolph (D-West Virginia), the senior Democrat on the Environment and Public Works Committee, saying that Gorsuch "is well qualified to administer the U.S. Environmental Protection Agency in a responsible and effective manner, and we urge your support of her confirmation."[15]

Along with the praise, however, came some harsh criticism, much of it from the environmental community. One-time chemical industry official and Delaware Republican Governor Russell W. Peterson wrote in his capacity as President of the National Audubon Society to "express dismay" over Gorsuch's nomination. In a letter to Senator Stafford, Peterson objected that Gorsuch "has no administrative experience or any technical background whatever to bring to the job of EPA Administrator—a job in crying need of such qualifications."[16] Peterson, whose credibility was enhanced by his having served as the Chairman of the Council on Environmental Quality under earlier Republican Administrations, said there are "two ways to undermine a government agency. One is by drastic budget cuts. The other is by the appointment of unqualified administrators. On both counts, the Administration is doing its best to render the Environmental Protection Agency ineffective."[17] Gorsuch's confirmation, Peterson contended, "will be an additional insult to the many millions of Americans dedicated to maintaining and improving environmental quality."[18]

Peterson's despair was shared by other environmentalists. Elizabeth A. Davenport, coordinator of Environmental Action, testified that Gorsuch "is an administrative novice and lacks necessary technical expertise."[19] "If someone with Mrs. Gorsuch's weak environmental credentials is confirmed, EPA will be unable to carry out its mandate. This could jeopardize environmental quality and the health and well-being of millions of Americans."[20]

The National Wildlife Federation, while not opposing the Gorsuch nomination, testified that it was "concerned" and that "there are questions about her effectiveness and leadership, which go to the very heart of EPA's independent status and of who will be making our national environmental policy."[21]

From her own state, the Colorado Open Space Council[22] also opposed her nomination. "We feel her depth of experience does not qualify her for this position," the group's Board of Directors said in a Western Union "Mailgram" to Senator Stafford.[23]

The early controversies over her qualifications and independence notwithstanding, Gorsuch won easy Senate confirmation, with just a few senators publicly expressing doubts and none voting against her confirmation. Once sworn into office, however, Mrs. Gorsuch quickly found the sledding to be rough.

It is generally agreed that a practical knowledge of how Washington works—and, importantly, of how and why it does not work much of the time—is crucial in administering an institution as complex as the Environmental

Protection Agency. That political "savvy" is equally important in avoiding the political land mines EPA faces in carrying out its far-reaching congressional mandate. It is helpful for top-ranking federal pollution control professionals, for instance, to have a personal understanding of the idiosyncrasies of the Congress, the Washington media, the industrial and public interest lobbying communities, and the internecine workings of the Executive Branch, particularly the Office of Management and Budget. It became apparent that the combination of having little substantive familiarity with EPA's programs and a seeming disdain for the inner-workings of Washington became major hurdles for Gorsuch and her agency to overcome.

One Washington "lesson" which Gorsuch seemed determined not to ignore was the tendency for some newly named agency officials to become "captives" of their agency's professional bureaucracies and of the inertia characterizing each particular agency. Gorsuch quite early in her tenure made it clear to agency professional staff and personnel that she was bent on *not* being overtaken by bureaucratic inertia, by tradition or precedent, or by overreliance on the agency's professionals. Although the EPA staff generally was regarded as among the most professional in the federal civil service, Gorsuch maintained a dividing line that separated her and a few key personal advisers from the rest of the agency's staff. The already existing physical barriers between the agency's "twelfth floor" Administrator/Deputy Administrator offices and the rest of the agency became even more impassable. Symbolism increasingly became reality.

Despite the Reagan Administration's overwhelming electoral mandate, the "us versus them" approach cost her early . . . and dearly. Ironically, it cost her in the loss from EPA of precisely the kind of talent which might have been most useful to her in the long run in administering the agency. Furthermore, the we/they mentality separating Administrator Gorsuch from the agency's career professionals created a situation in which lack of confidence and distrust soon flowed in both directions—with the inevitable result of reinforcing the mutual isolation—the captain from her troops.

Coming to EPA without the political stock to insist on selecting her own lieutenants—or at least having veto power over their selection by others, a common advantage for previous EPA administrators—Gorsuch was severely hampered even in dealings within the Administration. President Reagan had left no doubt that he would not be an apologist for environmental regulations, which he had frequently characterized as imposing unreasonable and unnecessary burdens on regulated industries. Throughout his campaign for President, Reagan stated his belief that many regulators had gone "too far" in seeking to protect the environment. EPA, the President's political supporters were convinced, had become over-staffed with activists bent on ignoring the need to "balance" environmental and economic programs. Yet, instead of appointing individuals qualified to achieve that balance, the Reagan Administration made it clear that EPA could become the dumping ground for political supporters unable to find a home elsewhere in government. This was no simple case of a successful candidate's using the traditional "spoils system" to fill top-level government appointments with qualified political allies. Nor was it a case of the newly elected President's filling a regulatory agency with qualified and competent administrators who could effectively carry out a popular mandate. Instead, the Administration's first round of appointees at EPA by and large consisted of individuals who, while loyal to President Reagan, simply lacked the depth of understanding and/or professional commitment necessary to fulfill their responsibilities.

Within five months of her confirmation by the Senate on May 20, 1981, the same Senate Environment and Public Works Committee which had voted unanimously to confirm her was convening oversight hearings to examine what Stafford, in characteristic understatement, referred to as "possible differences in opinion."[24] "One of the great problems the committee has encountered in the past few months is a lack of information as to happenings at the Environmental Protection Agency," Stafford told Gorsuch in welcoming her to the oversight hearings.[25] The Committee therefore had been forced to draw its impressions from a wide range of press accounts, and "the picture they paint is grim to a senator who has committed more than a decade of his career to laws intended to protect the sick, poor, and elderly by assuring them of healthful air and water."[26] "Morale is said to be the worst in agency history. Attrition is said to be 1 percent per month. The scene which these reports depict is of an agency in agony, with its senior officials under siege."[27]

Ranking Committee Democrat Jennings Randolph of West Virginia had instigated the oversight hearing with a September 30 letter to Stafford advising of "a proposed substantial reduction in funds and personnel for the agency at a time when additional responsibilities seem to indicate that resources should be increased or, at least, remain at existing levels."[28] "There are signs of organizational deterioration," he told Stafford, "which should not be permitted to develop further."[29]

Committee member Gary Hart (D-Colorado), who had unenthusiastically voted for Gorsuch's confirmation, summarized "persistent reports of disorganization at EPA" since Gorsuch had been confirmed:

- The recent resignations of two of the three Associate Administrators who were appointed to improve the agency's management;
- The failure so far of the Administration to nominate people to fill three of the six Assistant Administrator positions;
- EPA's proposed budget for fiscal year 1983, recently leaked to the press, which includes plans [to] fire more than 3,000 agency employees and to cut the agency's operating budget effectively in half;
- Widespread reports that policy at EPA is being made with no participation by the agency's professional staff; and
- Reports in the press that mass transfers of the agency's top career professional are being planned."[30]

"Taken as a whole, these items raise the fundamental question of whether EPA will be able to function effectively," Hart said.[31] It was a view that more and more legislators were expressing privately.

The controversies surrounding Gorsuch's management of the agency, over the performance of the agency's top political appointees, and over the nature of its legislative prescriptions intensified markedly during the succeeding year. Perhaps most controversial in the first year of her tenure was the Reagan Administration's handling of Clean Air Act amendments pending before the Congress. The Administration's initiatives satisfied virtually no interest involved in the volatile process of amending the Clean Air Act. After two sets of draft legislative proposals were leaked and roundly condemned on Capitol Hill, the Administration retreated and instead submitted a vague and generalized list of 11 "principles" which it hoped would guide Congress' deliberations. EPA's unimpressive and meandering performance in developing its position on the Clean Air Act legislation not only contributed to the congressional impasse already facing the amendments, but also raised widespread concerns over the competency of Gorsuch and her top officials in managing the agency's dealings with Congress.

In the early spring of 1982, 10 of the nation's most influential environmental organizations (not including the National Wildlife Federation) issued their most stinging report on environmental programs under the Reagan Administration. In *Indictment: The Case Against the Reagan Environmental Record*, the 10 organizations complained that "President Reagan has broken faith with the American people on environmental protection."[32] The President and his appointees, according to this viewpoint, "have simply refused to do the job that the laws require and that Americans expect of their government—to protect the public health from pollution and to use publicly owned resources and lands for the public good."[33] Directing their criticisms against a President whose election they had opposed in the first place, the groups said, "We think the Administration's environmental policies have harmed the Nation, and that the harm grows steadily worse. We believe President Reagan should be called to task."[34] Their complaints against EPA in particular spanned the entire range of agency programs—air pollution, hazardous wastes, water quality, toxic substances, enforcement, and research.

The environmentalists' criticisms were echoed and reinforced four months later, when three House subcommittees held an unusual joint hearing on EPA activities.[35] Testifying at those hearings, Senator Patrick Leahy (D-Vermont), who chaired the joint House and Senate Environmental Study Conference, concluded that "Ms. Gorsuch has so thoroughly mismanaged EPA's enforcement division that EPA's staff does not know what their job is or whether they dare to do it."[36]

Leahy pointed to frequent reorganizations of the agency's enforcement operations between July 1981 and May 1982, saying the reorganizations—at the rate of one every 11 weeks—had demoralized staff and had decimated the agency's enforcement record. Along with other critics, Leahy deplored the dramatic decline in EPA enforcement case referrals to the Department of Justice for prosecution.

In summarizing his view of Gorsuch's first year in office, Leahy told the three subcommittees: "[W]hen she was beginning the first of four reorganizations, she pledged that 'EPA is committed to a strong enforcement doctrine.' In spite of this pledge, mismanagement of the agency during the following months virtually destroyed its enforcement capability. In light of this record of mismanagement, there is little reason to assume that she will be sucessful in the future."[37]

Gorsuch's public defenders over time were becoming fewer and fewer, less and less willing to "stand up and be counted" on her behalf. Increasingly on the defensive in its dealings with the House and Senate, the Administration either withdrew its legislative initiatives or was forced by an unreceptive Congress to take a back seat in its substantive dealings on environmental legislation.

Nowhere, as it turned out, were the controversies more intense than with the hazardous waste programs, which had come to command the lion's share of press, public, and congressional scrutiny throughout the Gorsuch years at EPA. The individual chosen as President Reagan's nominee to run EPA's hazardous waste programs was Rita M. Lavelle. Like Gorsuch before her, Lavelle's nomination was controversial.

Lavelle, before being confirmed by the U.S. Senate to head EPA's hazardous and solid waste programs, had been a public relations officer with a chemical company subsidiary of the Aerojet General Corporation in California. From the start, she was controversial both because of her company's record of environmental and hazardous waste problems in California and, in particular, because of serious doubts about her intellectual and professional qualifications to serve as Assistant Administrator of EPA's Office of Solid Waste and Emergency Response. For critics, her resume alone provided an easy target: In listing her professional qualifications, Lavelle said she had "developed and implemented a major community relations campaign to counter environmental pollution charges brought against the corporation."[38]

Lavelle gained her entree to a presidential appointment through political contacts she had made in California as a press aide in the office of then-Governor Ronald Reagan. It was there, according to the authors of *A Season of Spoils*, that Lavelle came to know Edwin Meese, a long-time friend and political confidant of President Reagan. Meese held the Bible at Lavelle's swearing-in ceremony after her Senate confirmation as Assistant Administrator, and Anne Gorsuch has publicly acknowledged that she accepted the nomination of Lavelle only after her own protestations had been overruled.

Among the agency's political appointees and its professional staff and among congressional and outside interests, Lavelle soon came to be recognized as not measuring up to the job she had undertaken. Gary Dietrich, a top-ranking civil servant engineer during Lavelle's tenure, has been quoted as saying that accompanying her to testify before congressional committees was "the most embarrassing thing I ever had to do in the federal government."[39] Others who worked with her in the agency were no more charitable in evaluating her professional qualifications.

It was on Friday, February 4, 1983, that Administrator Gorsuch sought to remove Lavelle from office. Friction between the two had been apparent for some time, and Gorsuch's frustrations with Lavelle's management of the hazardous waste programs had been mounting. The immediate issue on February 4, however, was a memorandum apparently drafted by Lavelle to be sent to the White House. The draft memorandum suggested that Robert Perry, EPA's Associate Administrator for Legal and Enforcement Counsel, "is systematically alienating the primary constituents of this Administration, the business community."[40] When Lavelle refused to resign at Gorsuch's request, President Reagan three days later fired her. Later, with the discovery that paper shredders had just been installed in the agency's hazardous waste office, a kind of "feeding frenzy," as it has been described, was set off among the nation's media. EPA and Gorsuch and her top appointees became the prey, as headlines spread across the nation reporting one EPA controversy after another.

Lavelle later was convicted of perjury and obstructing a congressional investigation into the agency's handling of its hazardous waste management and clean-up responsibilities. She served four-and-a-half months of a six-month sentence in a minimum security "white collar" federal penitentiary before being released for good behavior.

Ironically, perhaps the most lasting impact of the Gorsuch-Lavelle controversies was on the legislative efforts in the 98th Congress, but not in the way the Reagan Administration had intended. Even in a "normal" era, EPA frequently is in a hot seat in its dealings with the Congress. While its programs can benefit a congressional district, they also can become very unpopular within particular districts, and among particular political constituencies such as regulated industries, municipal governments, or environmental organizations. EPA officials frequently point with some concern and disbelief to the large number—40 is an approximate consensus number—of House and Senate committees and subcommittees the agency must report and respond to in carrying out its responsibilities. Needless to say, effective congressional relations are an important aspect of the agency's success.

Few would question that effective working relations with the Congress—with both its Democratic *or* Republican members—was NOT an asset that the Gorsuch/Lavelle management team could call on in times of need. That became most clear when the House of Representatives on December 16, 1982, voted overwhelmingly to hold Gorsuch in contempt of Congress for the Administration's claim of Executive Privilege in withholding from a House subcommittee subpoenaed documents relating to Superfund cleanups.

In understanding Congress' approach to amending the Resource Conservation and Recovery Act in 1983 and 1984, it is essential that one be cognizant of the intense distrust of the Executive Branch that prevailed on Capitol Hill at the time. That highly charged "environment" set the tone for what became the Hazardous and Solid Waste Amendments of 1984.

Liquids in Landfills

Of the numerous disputes involving the Environmental Protection Agency during Anne Gorsuch's tenure as Administrator, perhaps no single controversy caused more damage to the agency than that involving its decision in the winter of 1982 to halt a scheduled prohibition on the disposal of containerized liquid wastes in landfills. The public, media, and congressional outcry provoked by the agency's suspension of a recently imposed regulatory ban on disposal of liquid hazardous wastes in landfills prompted the agency into an abrupt about-face within just four weeks. By then, however, the damage to EPA's credibility had been done, and the political fallout from the miscalculation, unlike the agency's decision itself, could not be readily reversed.

The train of events started in February 1982. Still learning the complex and inter-twining ropes of their new responsibilities, the agency's top management faced an upcoming prohibition on the landfilling of most containerized liquid hazardous wastes.

The regulations involved in the "liquids in landfills" controversy, as it became known, had been adopted by the Carter Administration's EPA in 1980. In May of that year, EPA had promulgated regulations applicable to owners and operators of hazardous waste treatment, storage, and disposal facilities operating with interim permits under RCRA Section 3005(e). Among the rules adopted by EPA was a prohibition on the disposal in landfills of most containerized liquid waste. The prohibition was to take place on November 19, 1980, but was uneventfully postponed several times. As described by the agency in the preamble to the regulations, EPA was concerned that containers disposed of in a landfill eventually would corrode and leak their liquid contents into surrounding soil and ultimately into groundwater.[41] The agency was concerned also that the containers of liquids, once the liquids escaped, would collapse and create voids. These voids would be filled by the subsidence from the surface of the landfill, thereby threatening the integrity of the landfill cap and exposing the landfill to infiltration from rainfall. The problem would be made worse because this series of events would occur over time, possibly only after an owner/operator had abandoned maintenance of the leachate collection system and the landfill cover.

Critics of the EPA scheduled ban on containerized liquids in landfills, in particular industrial representatives, objected to the ban in part on the basis that they needed more time if they were to design and construct the facilities and procedures for inspecting incoming containers and removing free liquids or solidifying those wastes. They also objected to the expense involved in eliminating the liquids from the waste drums. EPA, in early 1982, was planning to delay imposition of the ban as it had done previously, but it inadvertently failed to act. In a February 25, 1982, Federal Register notice EPA revoked the ban for 90 days and proposed other methods for dealing with the problem.[42] In so doing, however, the agency conceded the environmental problem stating that it "strongly believes that introduction of containerized free liquids in landfills should be minimized to the extent possible, if not prohibited." Further, the agency rejected the industry argument that more time was needed to prepare for the ban, noting that the regulated community had had 18 months to comply with the requirement, and that "some members of the regulated community have prepared to comply . . . ." As appropriate alternatives to landfilling liquids, EPA suggested incineration, deep well injection, solvent recovery and other forms of recycling as well as conventional wastewater treatment techniques. EPA also maintained that liquid and solid wastes could be separated at the point of generation, and that dewatering techniques or chemical/absorbent processes provided appropriate ways around the ban on landfilling of liquids.

However, in that February 1982 rulemaking EPA said it had "found merit" in criticisms over just what would constitute a "liquid waste" or a "waste containing free liquids." The agency was apparently convinced by critics' objections that no appropriate test protocols had been established for defining or measuring those properties.

In essence, EPA concluded that the prohibition on landfilling of liquid wastes was "too extreme for real-world application."[43] According to EPA, the prohibition, read literally, would mean that "landfill disposal of containerized wastes containing only 'one drop' of free liquid is banned. This would often require extraordinary, high-cost management practices to achieve compliance."[44] The agency pointed out that vibration and settlement as a result of shipment to a landfill would lead some small amounts of free liquids to separate from wastes in a container. Even with the best of efforts to adhere to the prohibition, the container might contain some free liquids. The landfill operator, if he were to assure compliance with the ban on disposal of "free liquids," would have to open and inspect all incoming containers, conduct some dewatering activities, and perhaps add some absorbents.

The agency concluded that "this opening, inspecting, and additional treatment operation by the landfill operator, in many instances, would add unnecessary costs and operational disruption and could present unnecessary personnel safety and environmental hazards because of the ignitability, volatility, or toxicity of many wastes commonly shipped in containers."[45]

All the same, EPA believed that the case for avoiding free liquids in hazardous waste landfills was strong enough to warrant some kind of "minimization" of free liquids in containerized wastes and was essential. The agency said it favored the kind of minimization "that could be achieved by reasonably simple and available dewatering practices and ordinary waste management practices."

So where did that leave EPA in its rulemaking? The agency acknowledged that it had no hard data for determining how much free liquids reduction might be achieved on a wide variety of hazardous wastes. Without being able to calculate how much "minimization" to require, it chose instead to rely on "professional judgment" in picking a figure.

The figure it was leaning toward at one point was 10 percent. That is, containerized wastes containing less than 10 percent of free liquids, by volume, would, for purposes of regulation and enforcement, be considered wastes having *no* free liquids. Ten percent wet . . . would be considered in practice to be dry. The result, EPA reasoned, would not only be practicable in the "real world," but also "a decided improvement over past practices in disposing of containerized wastes." The 10 percent free liquids criterion, in conjunction with other interim status standards, "would achieve reasonably acceptable environmental protection for interim status landfill operations."

Without justifying its optimism on the basis of research data, EPA concluded that although liquid wastes in landfills "often will eventually leak from their containers and migrate out of the landfill and into the environment," the agency nonetheless thought the leakage would be "slow, occuring over an extended period of years." Although such leakage of landfilled liquid wastes was inevitable over time, it "is likely to be considerably diluted and attenuated (in the environmental) [*sic.*]."[46]

The 10 percent free liquids threshold would reduce the amount of liquid that might leach. Other interim status closure and post-closure requirements would limit how much additional leachate generation would result from precipitation infiltration. And bulk or noncontainerized disposal of liquids would be regulated under still other requirements. As a result, EPA reasoned in February 1982, "the agency believes that any potential adverse environmental consequences will be substantially reduced."

The agency decided, however, against a literal application of the so-called "10 percent rule," because of other real-world considerations. The agency noted that in discussing its thinking with petitioners in ongoing *Shell Oil Co. v. EPA* litigation, which involved disposal of liquids in landfills, the agency came to appreciate that a 10 percent threshold approach still would leave some problems unresolved. Some containers still would have to be opened and inspected to assure compliance, and noncomplying containers still would have to have their free liquids absorbed. All the safety and environmental risks to landfill operators would remain. Furthermore, according to the petitioners in the pending litigation, getting certain wastes down to the 10 percent threshold still would require "extraordinary means of dewatering" in some cases. Couldn't the same ends be achieved by merely averaging the free liquid contents of "wet" wastes with other "dry" wastes, so that things in the end would average out? The petitioners essentially were asking whether some kind of "bubble policy" might apply in the liquids in landfills area just as EPA has decided the "bubble" can apply in averaging a source's air emissions and water discharges in some cases.

This proposal interested EPA. The agency decided in its February 1982 rulemaking to propose a rule it said would avoid the need to check each container for compliance with a numerical threshold. The agency determined that compliance should be based not on the free-liquid content of each individual container, but rather on the total number of containers having liquid wastes in a particular landfill.

The EPA proposal used a formula that would have allowed landfilling of drums containing liquids based on the depth of the particular landfill. The maximum amount of the landfill available for drums containing liquids was 25 percent, the figure for landfills 25 feet deep. For both deeper and more shallow landfills, less than 25 percent would be available for drums containing liquids. For instance, a 100-foot deep landfill would have been allowed to have no more than 10 percent of its contents devoted to drums containing liquids.

EPA acknowledged that its formula was derived from a proposal submitted to it by the National Solid Waste Management Association and the Chemical Manufacturers Association, parties in the *Shell Oil* litigation. Their proposal was intended to limit subsistence that would result from the collapse over time of drums containing liquids, rather than limiting the absolute amount of free liquids in landfills. Although the new approach, unlike the 10 percent per individual container rule, would not have restricted the amount of free liquids in individual containers, EPA maintained that "in actual operation," the "average container" would have *less* than 10 percent free liquid.[47]

Curiously, EPA appeared not to be convinced of its own reasoning. The proposed rulemaking declared that "the agency is concerned that today's amendment may not achieve the degree of minimization of containerized liquids in landfills that could be reasonably achieved because it does not directly limit the total amount of containerized free liquids placed in landfills. Consequently, EPA invites public comments on whether and how the amount of containerized free liquids allowed under today's amendment should be further limited."

While "dubious about the validity of such a technique," EPA specifically invited comment on the notion of allowing landfill operators to "tap" or "rock" a container and listen for the differences in sound between portions of containers filled with liquids and those portions containing solids." Another notion worthy of public comment was some kind of certification procedure under which generators would certify to landfill operators that the containers being delivered met standards for free liquid content.[48]

The Reaction . . . Immediate, Intense

If the EPA was having some difficulty determining whether its new approach might be satisfactorily protective of human health and the environment, others were having little such difficulty in forming their own opinions. Reaction to the 90-day suspension on enforcement of the liquids in landfills ban was intense, and generally extremely critical. The whole situation was turning into a political and public relations fiasco for an agency which was at that time just embarking on a full year of political turmoil. From a public relations standpoint alone, the most devastating image conveyed by the agency's decision not to proceed with the ban was that "hundreds of Love Canals" would be created. It was a graphic image for an American public that was already more than a little nervous about toxic waste problems. Naturally, critics of the EPA action milked it for all its worth.

"A wholesale retreat," complained Rep. James J. Florio, the New Jersey Democrat who chaired the key House subcommittee with jurisdiction over hazardous waste issues. Florio estimated that as much as 200,000 gallons per day of hazardous liquids would be disposed of in landfills as a result of the EPA decision not to enforce the ban. He predicted that the EPA announcement would undercut requirements that landfills have liners or leachate collection systems, discourage recycling and incineration of liquid wastes, and transfer to future generations the risks posed by today's unsafe disposal practices.[49]

The Chairman of the House Public Works and Transportation Investigations and Oversight Subcommittee, Georgia Democrat Elliott H. Levitas, labeled the EPA regulatory move "an incredible display of ignorance . . . . They are simply talking about poisoning American people. It is, indeed, an incredible act."[50]

To Republican Congressman Guy V. Molinari of New York, the EPA decision to put off enforcement of the liquids in landfills ban was a "reprehensible action and an insult to the public and the members of Congress." Within hours after the EPA action was announced, Molinari hypothesized to Levitas's subcommittee, "bulldozers were waiting" to dump toxic liquids into landfills not fit to hold them. "Even as we meet here," Molinari declared later at a public hearing, "the trucks are rolling into 900 landfills all over America carrying a deadly legacy that our children and grandchildren have no choice but to accept." He equated EPA's action with "opening all jails for a period of 90 days, to determine if the criminals within could possibly pose a threat to the public on the outside."[51]

For Congressman Marc L. Marks of Pennsylvania, the ranking Republican member of the Energy and Commerce Committee's Investigations Subcommittee, the EPA action was "an invitation to disaster," and one that "effectively removes all rules for the disposal of hazardous liquids." Declaring that the action was "patently and insultingly designed to give polluters a free ride at the expense of the health, safety, and piece of mind of the American public," Marks wondered if the EPA officials "making these reckless decisions" had taken the time to contemplate their potential personal liability. Marks suggested that certain EPA officials consider taking out administrative malpractice insurance: "They may very well need it."

Senator Edward M. Kennedy (D-Massachusetts) in a letter to EPA Administrator Gorsuch called the move "a dangerous and irresponsible signal . . .

that noncompliance with urgent health and safety measures will not be punished."[52]

The echoes of the controversy rippled across the nation's front pages and TV screens, as news of the suspension of the landfill ban spread across the nation. "U.S. Agency Seeks Easing of Rules for Waste Dumps," *The New York Times* headlined a March 1, 1982, story. "E.P.A. Wants to Allow Burial of Barrels of Liquid Wastes," the inside headline said.

"EPA Relaxes Hazardous Waste Rules" *Science* magazine headlined its April 16, 1982, article on the agency's action. "The agency gets bad marks from environmentalists, legislators, and the GAO," it continued in a large sub-headline.

"Battle Boils Over on Toxic Wastes," *U.S. News & World Report* headlined on March 22, 1982, adding that the EPA action had turned the nation's chemical waste dump sites and landfills "into an environmental battlefield between the Reagan Administration and its critics."

Even within the business community which generally had been viewed as the expected beneficiary of the EPA largesse, few were willing to step forward to publicly praise the agency decision to suspend the ban on landfilling of liquids. One of the nation's largest hazardous waste disposal companies, the IT Corporation, said it was pulling the plug on its activities aimed at developing regional treatment centers. A company vice president said IT was simply unwilling to invest more effort in the activity "unless and until there is a clear governmental commitment to the destruction and detoxification of hazardous wastes as the preferred method of disposal." Treatment industries standing to benefit by restrictive landfill rules joined with environmentalists in challenging the EPA rulemaking in court.

Pressed to respond by the kind of media and political frenzy that rarely greets a federal regulatory decision, EPA was forced to acknowledge its miscalculation. In an obvious effort at damage control aimed at halting the hemorrhaging it had inflicted upon itself, the agency on March 17, 1982, reversed its action announced less than one month earlier and reimposed the ban. "Given the hearing record and pending a final decision, I believe EPA should err on the side of caution," EPA Administrator Gorsuch said lamely.[53]

While the immediate liquids in landfills fiasco was over, the controversy it created was continuing to smolder. As much as any single public policy action taken in the environmental field since it took office in January 1981, the Reagan Administration's mishandling of the liquids in landfills issue precipitated a congressional backlash which not only played itself out in the 1984 Hazardous and Solid Waste Amendments to RCRA, but which continued to influence congressional attitudes in writing other pollution control legislation well into the middle of the decade.

To many—and particularly to those in the regulated community who continue to experience a backlash created by the liquids in landfills ban and other early mis-steps—the congressional reactions and, in some cases, over-reactions may amount to Anne Gorsuch's most lasting legacy to the environmental field.

Appendix: Are You Tough Enough? Anne Burford's "Insider's" Perspective

In her 1985 book *Are You Tough Enough?,* Anne M. Burford offered her own perspective on her nearly two years as Administrator of the Environmental Protection Agency. While self-serving in many respects, the McGraw-Hill book unquestionably provides the most detailed insight into her own attitudes toward her agency and those top officials with whom she served.

In terms of those responsible for the controversies that beseiged EPA during her tenure, Burford lays the blame broadly but squarely on Rita M. Lavelle, on the press corps, on the U.S. Congress, on the U.S. Department of Justice, on environmental organizations, and, ultimately, on President Reagan himself and his top White House staff.

On Rita Lavelle . . .

Rita Lavelle "was not my choice" to head the agency's highly visible hazardous waste management programs. Burford leaves little doubt on that point.

She says she initially declined to hire Lavelle "on the grounds that she lacked the background for this highly important, high-profile job." Her top staff's advice to "forget it" in terms of hiring Lavelle, however, fell on a deaf ear at the White House personnel office. "Nine parts persistence" is how Burford describes Joe Ryan, the White House personnel officer who repeatedly put Lavelle's name forward for the job of Assistant Administrator for the Office of Solid Waste and Emergency Response.

Ryan's "clincher," Burford writes, was his appeal to her on ground of sexism. She quotes Ryan:

"It becomes one of those Catch-22 situations: You don't have the management experience because you've never been given the opportunity for management experience, so therefore you're in this ever-downward cycle, and you as a woman should be sympathetic to that, and certainly you have been before."

Burford agreed to interview Lavelle, but "when I first saw Rita Lavelle, I was not favorably impressed. That may sound cruel, but it is the truth. The woman does not make a favorable physical impression. She is overweight, an unnatural blonde, and her appearance is blowsy."

Things got to the point, however, that "I was tired of looking, and the White House seemed to want her in the job." Partly because of her confidence in the top civil servants in the Office of Solid Waste and Emergency Response, she "gave in and agreed to hire Rita Lavelle."

Things went down hill quickly. "Almost from the beginning Rita was fighting over her turf, mainly trying to expand it, as opposed to being on the defensive. And she had a penchant for unhesitatingly bringing her problems right to me."

The situation did not improve. Burford complains that Lavelle conducted her own public relations activities separate from the agency's Office of Public Information. Lavelle's frequent and clandestine "contacts" with unspecified White House aides were a constant frustration to Burford. Lavelle's internal disputes with other EPA political appointees—and in particular with Robert Perry, Burford's enforcement chief—were a constant source of friction, as was Lavelle's refusal to be a "team player." Perhaps worst of all, Burford feels that Lavelle engaged in a pattern of lies and deceptions regarding her various activities.

Of Lavelle's perjury conviction and nine-month prison sentence, Burford writes that "It's possible that Rita Lavelle got off easy." She feels that had it not been for numerous "missing documents" (allegedly missing because Lavelle removed or destroyed them), the sentencing might have been more severe.

On President Reagan . . .

It is in criticizing the President that Burford is more delicate. When the President in 1981 asked her to head EPA, she writes, "I understood that he wanted me to carry out his policies of New Federalism and regulatory reform, and to get better environmental results with fewer people and less money." She writes that she accepted the President's nomination of her "because I wanted to bring a politically conservative approach to solving the management problems of environmental protection. . . . I thought Ronald Reagan shared that philosophy."

According to Burford, she came ultimately to believe otherwise. "Concluding that he doesn't care about the environment hurts," she writes. The President who "has always been a personal and political hero of mine" did not share what she says was her own commitment to environmental protection, and "having to face the fact that he does not is probably the hardest thing I have had to do."

"Frankly, it appears obvious to me that the President and his close advisors are simply ignoring the Environmental Protection Agency and its many mandates" from Congress, Burford concludes.

Burford's assessments of President Reagan's White House staff during her term at EPA are also harsh. Of White House Chief of Staff James Baker, frequently credited for many of the President's first-term legislative victories, she writes that "his understanding of environmental law was neither wide nor deep."

"He is supposed to be the great legislative strategist," Burford writes, "but I can recall any number of times during discussions of the Clean Air Act when his eyes glazed over repeatedly and came up peaches and oranges like a slot machine. (Eventually he shunted me off to an aide, Frank Hodsell, for which I was actually quite grateful because Frank not only knew the law, he was interested in it.) None of my many meetings with Jim Baker was satisfactory," Burford writes.

While she says she tried hard to keep Baker posted on EPA matters, "the man simply didn't understand them. Mr. Baker is *not* ignorant or ineducable. But as far as the environment was concerned, he didn't put forth any effort. It is clearly a secondary issue with him." Burford's concerns about Baker, however, went beyond her mere feeling that he did not *understand* environmental issues. She felt also that Baker (now Secretary of the Treasury) simply did not *care* about protecting the environment. "Another point in Hodsell's favor," she writes in contrast to Baker, "was that not only did he understand, he also seemed to care about environmental matters."

Her view of another of the President's top aides, Craig Fuller, is also unequivocal. She once liked Fuller "a great deal," she writes. But after later discovering that he was a frequent confidential White House contact for Rita Lavelle, "I have to lump him along with all the others I've met in Washington who follow a private agenda."

On the U.S. Congress . . .

In one of her more widely-quoted remarks from *Are You Tough Enough?,* Burford characterizes her experiences as EPA Administrator as "an expensive mid-life education." She asks: "But was it an education I needed to acquire?" She thinks not, reasoning that "In some ways I would have preferred my illusions."

Among the "truths I came to see" Burford lists—

—Congress' "lack of concern with and its lack of accountability for the terribly serious budgetary problems of the country." She complains of Congress's "dereliction of its budgetary duty" in its approving budgets higher than requested by the Reagan Administration for all programs other than defense.

—Congress's preoccupation (and also the media's) with issues such as the number of employees EPA has and the size of its budget, "not with the quality of the job we were doing."

—While Burford has "no criticism" when it comes to her dealings with House Energy and Commerce Committee Chairman John D. Dingell, Michigan Democrat, other unnamed members of that committee, she complains, were "men without integrity, men who only sought to abuse and get headlines."

On the U.S. Department of Justice . . .

Anne Burford reserves her harshest criticisms for the Department of Justice and those who headed it during her tenure at EPA. Her criticisms apply primarily to Justice's handling of her defense when in December 1982 she was cited by the House of Representatives for contempt.

Justice's lawsuit defending her in her claim of Executive Privilege to withhold "enforcement sensitive" documents from the Congress was "the sloppiest piece of legal work I had seen in 20 years of being a lawyer," she writes. "It was just plain sloppy work. It was disorganized, it had simple facts wrong, and it even contained misspellings. There were also fundamental mistakes." By Burford's account (an account subsequently substantiated at least in part by a lengthy investigatory report by the House Judiciary Committee), Justice's top attorneys under Attorney General William French Smith repeatedly excluded her from negotiations and planning sessions involving her own legal defense.

At one point, Burford virtually accuses Justice of conducting its own cover-up while in effect hanging her out to dry. Regarding a nonsensitive document requested by the Congress as part of its investigations of the hazardous waste program, Burford writes that Justice "conveniently mislaid an important piece of evidence that would have brought embarrassment, or worse, to some of its own people."

While not directly criticizing Attorney General William French Smith, Burford relates a line from Stanley Brand, who was Counsel to the Clerk of the House of Representatives, the attorney representing the House in its contempt of Congress citation against her.

"He was a part-time Attorney General," she quotes Brand as saying. " 'If you wanted to hide something from Mr. Smith, put it in a law book.' "

On the Media . . .

"One central fact should be understood," Burford writes unequivocally. "In the almost nine years that had elapsed since the unraveling of the Watergate case, no Washington story received as much coverage as the 'Scandal' at EPA. . . . Except for Patty Hearst's, I can't recall any story anywhere in the United States that received as much attention.

Burford may have been struck by the quantity of press coverage she received, but she clearly was less impressed with the *quality* of that coverage. With some justification, Burford complains of "pack journalism" and of unattributed and unsubstantiated accusations. But her broad brush approach is broad indeed when it comes to her assessment of the nation's media covering environmental issues:

"As for the press, I am told that there *are* good, honest journalists who do their homework, get their facts straight, and keep their biases to themselves," she writes. "Unfortunately, I have met only a very few of them. The vast majority, in my opinion, do not work hard and let opinion into their news copy on a regular basis."

Once reporters allow their personal opinions to enter their stories, Burford leaves little doubt that she knows precisely where those opinions will fall: She refers at one point to "the liberal press (if that isn't a redundancy)."

"The media, the leftist media, intentionally distorted this issue against me," she complains at another point.

"At the risk of sounding paranoid," Burford writes, "let me state that our good work was not properly reported because the press brought the charges and the rhetoric of the environmentalists without checking for substance."

"It both scares and sickens me to realize that *policy* of the United States government can be determined by the front page of *The Washington Post,*" she writes. Without detailing the "it" of which she complains, she writes: "If they did it at EPA, they can do it again. Perhaps this time at the Department of Defense? One of the things that my case so clearly illuminates is that the press prefers to cover personalities rather than issues, and as a result public trust, which is essential to good government, does not rest on fact and reality. It can so easily be manipulated by allegation and distortion. I really feel that the time has come for the press to clean its own house."

Washington-based reporters care "only for how many people you have and how much money you have—the size of your turf," according to Burford. "Forget about results."

And, At Last, on Washington . . .

"This is a funny town, Washington. Or had you already figured that out?" Burford asks rhetorically near the end of her book.

She describes what happened to her in that "funny town" as being "scandalous." But the neglect of serious issues at EPA from others in Washington "is worse."

Referring to herself by a nickname bestowed on her by her more virulent critics, Burford writes: "Now, with the Ice Queen gone, no one seems to be paying any attention. No one seems to be standing up and asking the right questions. Maybe no one is tough enough."

Chapter 2 Footnotes

[1] "Nominations of Anne M. Gorsuch and John W. Hernandez, Jr.," (henceforth "Gorsuch Nomination"), Hearings before the Committee on Environment and Public Works, U.S. Senate, 97th Cong., 1st Sess., pp.277-280, May 1 and 4, 1981 (Serial No. 97-H26).

[2] *Id.*

[3] *Id.*

[4] *Id.*

[5] "Gorsuch Nomination," 32.

[6] *Id.*

[7] *Id.*

[8] "Gorsuch Nomination," 33.

[9] "Gorsuch Nomination," 34.

[10] *Id.*

[11] *Id.*

[12] "Gorsuch Nomination," 25.

[13] Letter of April 30, 1981, from William A. Cox, Jr., P.E., to Senator Robert T. Stafford, reprinted in "Gorsuch Nomination," 178.

[14] Letter of March 30, 1981, from Denver Mayor William H. McNichols, Jr., to Senator Robert T. Stafford, reprinted in "Gorsuch Nomination," 173.

[15] Letter of April 7, 1981, from Hubert Beatty to Senator Jennings Randolph, reprinted in "Gorsuch Nomination," 167.

[16] Letter of May 1, 1981, from Russell W. Peterson to Senator Robert T. Stafford, reprinted in "Gorsuch Nomination," 189.

[17] *Id.*

[18] *Id.*

[19] "Gorsuch Nomination," 66.

[20] *Id.*

[21] "Gorsuch Nomination," 65.

[22] *Reagan's Ruling Class: Portraits of the President's Top One Hundred Officials*, Brownstein, Ronald, and Easton, Nina, Presidential Accountability Group, Washington, D.C., 1982, p.207. (The Colorado Open Space Council's views of Gorsuch as a state legislator improved significantly in her second term compared with her first. She earned 33 and 8 percent "favorable" ratings in 1977 and 1978 and 73 and 72 percent ratings in 1979 and 1980. There are some who question whether the improved ratings reflect more a change in the council's rating procedures than an improvement in Gorsuch's approach to environmental issues per se.)

[23] "Gorsuch Nomination," 185.

[24] "Environmental Protection Agency Oversight," (henceforth "EPA Oversight"), Hearing Before the Committee on Environment and Public Works, U.S. Senate, 97th Cong., 1st Sess., 1, October 15, 1981, (Serial No. 97-H29).

[25] *Id.*

[26] "EPA Oversight," 2.

[27] *Id.*

[28] "EPA Oversight," 4

[29] *Id.*

[30] "EPA Oversight," 10.

[31] *Id.*

[32] *Indictment: The Case Against the Reagan Environmental Record*, March 1982, 1.

[33] *Id.*

[34] *Id.*, 2.

[35] "Joint Hearings," before Certain Subcommittees of the Committee on Government Operations, the Committee on Energy and Commerce, and the Committee on Science and Technology, U.S. House of Representatives, 97th Cong., 2nd Sess., 29, July 21 and 22, 1982 (Serial No. 970199), 29.

[36] "Joint Hearings," 45.

[37] "Joint Hearings," 46.

[38] *A Season of Spoils*, J. Lash, K. Gillman, D. Sheridan, Pantheon Books, New York, N.Y., 1984, p.42.

[39] *Id.*, at 42.

[40] *Id.*, at 77.

[41] 45 Fed. Reg. 33066 (May 19, 1980).

[42] 47 Fed. Reg. 8307.

[43] 47 Fed. Reg. 8308 (February 25, 1982).

[44] *Id.*

[45] *Id.*

[46] *Id.*

[47] 47 Fed. Reg. 8309 (February 25, 1982).

[48] *Id.*

[49] *Environment Reporter-Current Developments*, BNA, March 5, 1982, p.1388.

[50] *Id.*

[51] Environmental Protection Agency public hearing held in Washington, D.C., March 11, 1982.

[52] Letter addressed to Environmental Protection Agency Administrator Anne M. Gorsuch from Senator Edward M. Kennedy, dated March 5, 1982.

[53] *Environment Reporter-Current Developments*, BNA, March 19, 1982, p.1476.

Chapter 3

Common Law

Just as the 1984 RCRA Amendments have a political context, they also have a legal one. It consists of the common law and the previous statutory efforts to control hazardous waste.

Common law—law made by judges—is the precursor not only of hazardous waste control statutes but of pollution control laws generally.[1] Although environmental protection now is governed principally by statutes and regulations, common law has played a significant and useful role, including, in some instances, providing compensation for those injured by exposure to hazardous substances.* In a few cases it has been used successfully in forcing the closure and cleanup of hazardous waste sites.[2] The purpose of this chapter is to discuss the principal common law doctrines that have been used in environmental litigation and describe how their limitations became a factor in increased reliance on statutes, such as the Resource Conservation and Recovery Act, to accomplish the goal of environmental protection.

Perhaps the most familiar common law doctrine in environmental cases is that of nuisance. The term nuisance, derived from the French word for harm, actually encompasses two concepts. A private nuisance is an activity or condition which creates an unreasonable interference with the use or enjoyment of private property.[3] A public nuisance usually involves an unreasonable danger to public health or safety.[4] Lawsuits invoking the doctrine of nuisance (both public and private) have been used to challenge nearly every type of industrial activity—from textile mills to oil refineries to quarries and mines.[5] Over the years, nuisance actions frequently have targeted pollution from waste materials.[6]

The question of whether a nuisance should be eliminated or restricted involves a balancing process in which the gravity of the harm is weighed against the utility of the defendant's activity.[7] Balancing tests by nature are subjective, and the case law yields a bountiful collection of inconsistent results. For example, in *Madison v. Ducktown Sulfur, Copper & Iron Co.*[8] several farmers sued to enjoin the smelting operations of two companies that used open air "roast piles" which created large quantities of thick smoke that damaged the farmers' trees and crops. The Tennessee Supreme Court approved the award of compensation but refused to grant the injunction:

> In order to protect by injunction several small tracts of land [worth less than $1,000] we are asked to destroy other property worth nearly $2,000,000 and wreck two great mining and manu-

*For example, in *Hagy v. Allied Chemical & Dye Corp.*, 122 Cal. App. 2d 361, 265 P.2d 86 (1953), defendant's creation of a sulphuric acid fog was held to be the proximate cause in aggravating plaintiff's dormant cancer. More recent examples of "toxic tort" litigation involve attempts to recover for exposure to asbestos and Agent Orange. Problems encountered in this type of litigation have generated efforts to enact "victims compensation" legislation. See, T. Garrett, "Compensating Victims of Toxic Substances: Issues Concerning Proposed Legislation," 13 ELR 10172; Zazzali and Grad, "Hazardous Wastes: New Rights and Remedies? The Report and Recommendations of the Superfund Study Group," 13 *Seton Hall Law Review* 446 (1983); Environmental Law Institute Report, "Toxic Tort Litigation: An Analysis of Pending and Terminated Hazardous Chemical Personnel Injury Cases in Massachusetts Federal and State Courts," (Sept. 1984).

> facturing enterprises . . . and to deprive thousands of working people of their homes and livelihood.[9]

A contrary approach was articulated by the California Supreme Court in *Hulbert v. California Portland Cement Co.*[10] In that case, the court granted an injunction to protect citrus fruit growers from damage caused by the defendant's cement dust. According to the court, "If the smaller interest must yield to the larger, all small property rights and all small, less important enterprises . . . would sooner or later be absorbed by the large, more powerful few . . . "[11]

In addition to inconsistent results, nuisance suits are characterized by the denial of relief where the damage is speculative or uncertain. Thus, it is difficult to abate a substantial environmental risk that has not yet caused substantial damage.[12] In most nuisance cases involving active businesses, courts take into account such factors as investment and employment, along with environmental harm. While damages are sometimes awarded, the closure of the polluting facility is an unusual remedy. The polluter may be required, however, to implement the "best available technology" to control, if not eliminate, the effects of the nuisance.[13] This common law standard has been incorporated into several environmental statutes, including the Clean Water Act.[14]

Relatively few nuisance cases have considered the problem of groundwater contamination; those that have, as with nuisance generally, have produced inconsistent results. In an English case, decided in 1885, the defendant used an abandoned water well on his property for the disposal of sewage. The sewage "percolated through the underground strata" and contaminated the plaintiff's well water. The court held that the defendant had not put his well to its natural and proper use. The rule of law, according to the court, is: "Though a man may suck dry the stratum which is a common sponge for himself and his neighbors, he is not entitled to poison the sponge."[15]

Across the Atlantic at about the same time, the Florida Supreme Court, deciding the case of *Pensacola Gas Co. v. Pebley*[16], also concluded that groundwater contamination could constitute a nuisance. There a homeowner sued a manufacturer of "illuminating gas" located two blocks from his residence. The court cited evidence which showed that the manufacturer had allowed "yellowish water and tar water" to "run upon the ground and to penetrate same so that . . . in the spring of 1885 the plaintiff's well became polluted, and the water tasted and smelt like gas." The water "became so unpalatable, nauseous and unhealthy that it was unfit for drinking or bathing."[17] The court ruled that the defendants "had no right to allow the filthy water to escape from their premises and to enter the land of their neighbors."[18]

In contrast, in a New York case,[19] it was held that an oil refinery that discharged large quantitites of process water "impregnated with chemicals and acids" onto the ground was not liable for the resulting groundwater contamination. According to the court, "it is only in exceptional cases that the owner can know beforehand that his works will affect his neighbors wells . . . [thus in the absence] of negligence or knowledge as to the existence of subterranean water courses, when the business is legitimate, there can be no liability if [the groundwater] becomes contaminated.''[20]

The Court of Appeals of Kentucky took a different view in the case of *Kinnaird v. Standard Oil Co.*[21] The court ruled that a nuisance was created as the result of the storage of coal oil ''that had leaked from the casks and being of such penetrating character, passed into the ground and polluted the water'' supplying the plaintiff's spring. The court held that regardless of whether the defendant was negligent, ''he had no right to pollute the water in such a manner as when it passes to his neighbor its use becomes dangerous or unhealthy to his family or to the beast on his farm.''[22]

Trespass, another common law doctrine, has not proven as useful in environmental law as nuisance because it can be employed only where private property has been invaded by some physical or observable object which the defendant set in motion.[23] As a general matter, a successful trespass action must be based on proof that the intrusion was substantial and that it invaded a legally protected interest in property.[24] Although a property invasion is common in hazardous waste cases, usually in the form of leaks and other releases, trespass actions may be brought only by those who own or are in possession of the affected property.[25]

In earlier common law cases, a major advantage for the plaintiff in using a trespass action was that proof of negligence was not required, simply an invasion of property.[26] During the twentieth century, however, the trend has been to require evidence either that the trespass was the result of an ultra hazardous activity, or that the defendant acted intentionally or negligently.[27] For example, in one case, the unintentional, non-negligent seepage of chemicals from the defendant's facility which resulted in contamination of plaintiff's well water was held not to constitute a trespass.[28] Nevertheless, the modern trespass action offers at least one significant advantage. Once a court finds that a trespass has occurred, it usually will not engage in a balancing test, used in nuisance cases, of weighing the plaintiff's injury against the utility of the defendant's enterprise.[29] It should be noted, however, that the Oregon Supreme Court has established a test for trespass that, as a practical matter, cannot be distinguished from the test for nuisance.[30]

Another common law doctrine pertinent to hazardous waste litigation is strict liability for conditions and activities whose risk of harm cannot be eliminated by the exercise of due care. An action brought under this doctrine requires proof that *under the circumstances* the condition or activity qualifies as ''abnormally dangerous'' and was the proximate cause of plaintiff's harm.[31] The doctrine of strict liability for ''abnormally dangerous activities'' derives, in large measure, from a famous 19th century English case, *Rylands v. Fletcher*. The defendants in *Rylands* were mill owners who had constructed a reservoir on their land. The bottom of the reservoir failed to hold, and the escaping water flooded the plaintiff's coal mine. Although the defendants were found not to have been negligent in the construction of their reservoir, the House of Lords affirmed the judgment of the Exchequer Chamber, which held:

> [T]he person who for his own purposes brings on his lands and collects and keeps there anything likely to do mischief if it escapes, must keep it at his peril and if he does not do so is *prima facie* answerable for all the damage which is the natural consequence of its escape.[32]

Adapting the *Rylands* case to the escape of hazardous waste, the court in *New Jersey v. Ventron Corporation* ruled that a series of owners of a mercury processing plant were responsible for the pollution of an adjacent creek based on the "common law principle of strict liability for unleashing a dangerous substance during non-natural use of land."[33] Employing the same theory of strict liability, the Supreme Court of Utah held that placing toxic waste water adjacent to the plaintiff's water wells constituted an unreasonably dangerous and inappropriate use of the defendant's land. The court rejected the defendant's contention that since it was engaged in a legitimate business activity, a negligence test should have been applied.[34]

Although much more difficult to prove than strict liability cases, tort actions based on negligence have also been used extensively by private parties in environmental lawsuits. Negligence is generally defined as conduct which falls below the standard established by law for protection against unreasonable risk of harm.[35] Unlike an early common law action based on trespass, which theoretically required no proof of actual harm to the plaintiff, a successful negligence action requires that a plaintiff prove actual injury and that the defendant's conduct was the proximate cause of the injury.[36] In *O'Leary v. Moyers Landfills, Inc.*, the court recognized that the leachate from the defendant's dump site contained known and suspected carcinogens, but held that under the law of Pennsylvania the plaintiffs had not proven actual injury as a result of leachate discharges, and subsequently dismissed the negligence claim.[37] Perhaps the principal difficulty with relying on a negligence claim in a hazardous waste case is demonstrating that the defendant violated his standard of care. This can present a substantial burden of proof because, due to the general lack of knowledge about the problems associated with hazardous waste, the duty of care that is required is difficult to define. For instance, in *Ewell v. Petro Processors of Louisiana, Inc.*,[38] the generators of hazardous waste were found not liable because they neither knew, nor had reason to know, that their wastes were leaking from the landfill where they had been disposed of.

Similar reasoning was used by the federal district court in *Wright v. Masonite Corporation*. In that case, the court held that under North Carolina law formaldehyde gas that was released from the defendant's plant and damaged food in the plaintiff's supermarket was not a nuisance because it had not been established that the plant was operated improperly. However, other courts have held that those who transport or dispose of potentially harmful substances have a fundamental legal obligation to protect other persons, and their property, from unreasonable risk of harm.[39]

In addition to the problem of inconsistent results (see following section entitled "Tale of Two Dumps") several other obstacles have prevented common law actions from becoming the vanguard in the battle against environmental degradation. First, in private litigation, the law has been principally concerned with monetary compensation for the individual litigant, rather than protection for the public generally. Second, litigation is expensive and consequently a sufficient number of cases cannot be brought to reach many environmental problems. Third, lawsuits, by their very nature, usually are brought after the damage has occurred; very rarely has private litigation served primarily as a means to prevent environmental damage. Finally, the more difficult the problem is, in terms of its technological complexity, the greater the motivation for judges to transfer the problem to state and federal agencies that possess the pertinent expertise.

Thus, with a gradually developing consensus that courts and common law rules could not adequately address widespread environmental pollution, citizens and their elected representatives endeavored to find legislative solutions. The results by 1986 comprise a vast and complex network of federal and state statutes and regulations—as well as local and regional rules.

Yet, despite its limitations, it would be premature to conclude that the role of common law in hazardous waste cases has been wholly displaced by statutes and regulations. In 1979 and 1980 the Department of Justice, in order to establish the defendant's liability for acts occurring before RCRA's enactment in 1976, generally included a common law public nuisance allegation when it filed a RCRA Section 7003 imminent and substantial endangerment case. Prior to the enactment of Superfund on December 10, 1980, the Department filed more than 50 such cases.[40] Although the liability provisions of Superfund seemed to preclude the necessity of including an allegation of common law nuisance, it should be recalled that the Superfund law itself not only is premised on common law principles, but specifically preserves liability under common law.[41] Long-standing principles of common law are intimately related to the government's Superfund enforcement efforts in situations where numerous companies have contributed wastes to a site. The doctrine of joint and several liability holds that where two or more tortfeasors have contributed to a single *indivisible* injury, each defendant may be held liable for *all* of the damages. In the hands of the government's Superfund litigators, the common law doctrine of joint and several liability has been a powerful and intimidating weapon.[42]

In RCRA, a statute in which the liability provisions in the 1984 Amendments are fairly comprehensive, the need for the direct application of common law principles is less apparent. Nevertheless, the legislative history of the 1984 Amendments provides substantial evidence that Congress did not consider RCRA (even with the extensive additions) to be "complete." Thus, in Congress' view, common law should be kept in reserve and used when needed.

According to Congressman James J. Florio (D-New Jersey):

> Congress cannot reasonably be expected to anticipate every possible aspect of enforcement litigation. . . . It would be ab-

> surd to arbitrarily preclude the Government from resorting to common law principles where appropriate. It should be kept in mind that Section 7003, which authorizes lawsuits to force cleanups of hazardous waste sites is founded on long standing principles of common law nuisance.[43]

Senator John Chafee (R-Rhode Island), the floor manager for the RCRA reauthorization bill in the Senate, agreed: "We must continue to rely to a large extent on Federal and State common law and State statutory law to fill the gaps."[44]

Appendix: A Tale of Two Dumps

The use of common law to address hazardous waste sites can be seen in two lawsuits from states a half-continent apart, involving the land disposal of massive quantities of chemical wastes.

The Pig Farm in Coventry, Rhode Island[45]

Piggy Hill Lane is a winding dirt road on the outskirts of Coventry, Rhode Island, that leads directly to a five-acre clearing owned by the Picillo family. Near the entrance of their property, the Picillos raised pigs—but that wasn't their only visible means of support. Less than 1,000 feet away was a trench about 200 feet long, 15 to 30 feet wide, and 15 to 20 feet deep. The periphery of the trench was lined with more than 100 fifty-five gallon drums and five-gallon containers. Covering the bottom of the trench was a layer of "pungent, vari-colored liquid" to a depth of six inches at its shallowest point. The method of disposal was fairly straightforward: a truck driver with a load of chemical waste would back a truck up to the trench and then push barrels off the tailgate. The chemicals drained from the damaged barrels into the trench.

Apparently the Picillos' fire prevention system achieved approximately the same level of sophistication. On September 30, 1977, the trench exploded. Firefighters responded, but they were unable to extinguish the fifty-foot flames. According to witnesses, as the fire raged within the trench, several more explosions were heard, and clouds of thick smoke blackened the eastern horizon.

Immediately after the fire, the Picillos were ordered by the state fire marshall to cease all disposal operations and remove all flammable wastes. The orders were ignored. Consequently, the Rhode Island Attorney General and several of the Picillos' neighbors filed a lawsuit charging the Picillos with maintaining a public and private nuisance.

At the trial, neighbors testified that during the summer of 1977 pungent odors frequently had forced them to stay inside their homes. One neighbor complained that the fumes from the trench gave her headaches and made her nauseous. Another neighbor testified that on one occasion the odors made her cough severely, and gave her a sore throat which lasted for days.

Another witness, Dr. Nelson Fausto, a professor and pathologist at Brown University's Department of Biological and Medical Science, described the potential health effects of some of the chemicals discovered in the trench. Chloroform, a narcotic and anesthetic, he testified, would induce vomiting, dizziness and headaches in persons exposed to it. Trichloroethane and trichloroethylene, according to Dr. Fausto, have chemical structures that are similar to chloroform and produce similar toxic effects. These three chemicals are strong carcinogens that cause hepatoma (cancer of the liver). Dr. Fausto testified also that toluene and xylene are toxic, and that exposure can cause irritation of the mucous membranes in the upper respiratory tract. In addition, chloroform that is heated to sixty-eight degrees Fahrenheit converts to phosgene gas—the toxic nerve gas used in World War I. Direct sunlight, according to Dr. Fausto, would be sufficient to produce phosgene gas from the chloroform present in surface water.

Another expert witness, William Kelly, Ph.D., an Associate Professor of Civil and Environmental Engineering at the University of Rhode Island, testified that the permeable nature of the soil permitted the chemicals in the trench to percolate down to the water table and travel with the groundwater northward at a rate of one foot per day into a marsh. Waters from the marsh drain into several ponds and the Quinabog River.

Relying on these facts, the trial judge found that the Picillo's dump site constituted both a public and private nuisance. He issued a permanent injunction against disposal operations and ordered that all chemicals and contaminated soil be taken away and disposed of at a permitted disposal facility.

The defendants appealed the judge's ruling, contending that the plaintiffs had failed to prove that the Picillos had been negligent in disposing of hazardous waste on their property. On April 9, 1982, the Rhode Island Supreme Court unanimously rejected the Picillos' defense. The court held that liability for a nuisance in Rhode Island is predicated on "unreasonable injury rather than on reasonable conduct." Therefore, "negligence is not a necessary element of a nuisance case involving contamination of public or private waters by pollutants percolating through the soil . . . "

In reaching its decision, the court was obliged to address a 1934 case, *Rose v. Socony Vacuum Corp.*,[46] cited by the defendants. In that decision, the Rhode Island Supreme Court had ruled that the defendant could with impunity contaminate the plaintiffs' drinking water if the defendant polluted non-negligently. The court had reasoned that because the "courses of subterranean water are . . . indefinite and obscure," it would be unfair to impose liability on land owners for the unforseeable consequences of legitimate land uses.

In dealing with the precedent in *Rose*, the court in the *Picollo* case did not attempt to distinguish it on its facts. Instead, the court pointed out that since 1934, "the science of groundwater hydrology and societal concern for environmental protection has developed dramatically [and] the courses of subterranean waters are no longer obscure and mysterious."[47]

The court also observed:

> [D]ecades of unrestricted emptying of industrial effluent into the earth's atmosphere and waterways has rendered oceans, lakes, and rivers unfit for swimming and fishing, rain acidic, and air unhealthy. Concern for the preservation of an often precarious ecological balance, impelled by the spectre of "a silent spring," has today reached a zenith of intense significance.[48]

Consequently the court found that, "the scientific and policy considerations" which led to the decision in the *Rose* case, "are no longer valid." The Picillos' appeal was "denied and dismissed."

Devil's Swamp in Louisiana[49]

The Ewell family owned an eleven-hundred-acre tract of land between Scenic Highway and the Mississippi River, north of Baton Rouge, Louisiana. About half the tract is marshland and part of an area known as Devil's Swamp. The Baton Rouge Bayou flows into Devil's Swamp from the northeast, close to the north end of the Ewell's property. The adjacent land to the north was owned by Petro Processors, Inc., a company in the business of hazardous waste disposal.

Petro Processors' customers, industrial plants in the Baton Rouge area, contracted to have their wastes picked up and disposed of. These wastes were either buried or dumped into large pits on Petro Processors' property. In June 1968, three new pits were dug about three-hundred feet north of the Ewells' property line. About eighteen months later, the waste material from at least one of these pits began to leak onto the Ewell property, contaminating the 550 acres of marshland with chlorinated hydrocarbons. In some areas, the concentration was as much as 25,000 parts per million.

The Ewell family sued Petro Processors and 10 of its customers: Humble Oil & Refining Co., the Dow Chemical Co., Ethyl Corporation, Allied Chemical, Rubicon Chemicals, Inc., Enjay Chemical Company, Foster Grant Company, Uniroyal, Inc., Copolymer Chemical Company, and Shell Chemical, Inc. Prior to trial, most of the Ewell family reached an out-of-court settlement with the defendants, but Dave Haas Ewell, Jr., and his children, who owned a one-eighth undivided interest in the property, refused to settle. The case was tried before a jury, which held all of the defendants responsible and awarded the plaintiffs $25,000 for property damages and $5,000 for mental anguish. The plaintiffs, who wanted substantially more compensation and also a comprehensive clean-up of their property, appealed.

On October 9, 1978, the Louisiana Court of Appeals handed down its decision. The court held that there was no doubt that Petro Processors had negligently permitted toxic wastes to contaminate the Ewells' property and therefore was liable for any damages resulting from its negligence. However, "[a] more difficult question is presented as to the other defendants." The court's analysis was premised on the finding that Petro Processors was an independent contractor whose customers provided no instructions concerning the method of disposal.

According to the court, an employer is not liable for the offenses of an independent contractor. An exception to this general rule "is that if the work is inherently or intrinsically dangerous, . . . the employer cannot avoid liability by letting the work out to an independent contractor." However, if such work can be done safely, the employer cannot be held liable. The court then ruled:

> [T]he record in this case does not support the conclusion that the work done by Petro Processors cannot be done safely. The damages suffered by plaintiffs were due, not to the dumping of toxic substances in the pit, but to the improper construction of

> one of those pits. We therefore conclude that the customer defendants cannot be held liable unless it is shown that they were aware of the leakage at the pits and continued to dump hazardous material at that site.[50]

The court reviewed the record of the customer defendants, their knowledge of the disposal facility, and the type of chemicals disposed. According to its analysis, the record exonerated all but one of the customer defendants.

> Ethyl is the only customer defendant which is shown to have dumped the pollutant in question from before the pits were found to be leaking until well after this suit was filed. Since it admitted that several of its employees visited the pits, knowledge of the condition thereof can certainly be attributed to it. We therefore find that Ethyl is liable solidarily with Petro Processors for the damages done to plaintiffs' property.[51]

The court then turned to "the major question presented by this case"—the measure of damages. The plaintiffs argued that money alone could not cure the damage, and that those liable should be responsible for restoring the environment to its original condition.

The court disagreed, pointing out that the "property is a swamp, subject to periodic overflow from the Mississippi River and the Baton Rouge Bayou." In addition, the court stressed the property's limited commercial value and the enormous cost of clean-up.

> Prior to being polluted, it was used seasonably for grazing of cattle, and non-commercially, for hunting and fishing. Its only other commercial use was for the growing of timber. Its value was set at $375.00 per acre, or slightly over $200,000 for the 550 acres affected, by the only expert witness who testified as to that point. The restoration of the property, according to plaintiff's witness, would take about seven years, involve the use of 100 trucks running continuously during that time, and would cost 170 million dollars.[h]

The court ruled that, "under those circumstances, the proper measure of damages is the diminution in value of the affected property."[i] Thus the judgment of the trial court was affirmed except that the only defendants held liable for the $30,000 award were Petro Processors and Ethyl Corporation.

Chapter 3 Footnotes

[1] See W. Rodgers, *Environmental Law,* 100-163 (West Publishing Co. 1977) and F. Anderson *et al., Environmental Protection: Law and Policy,* 56-59, 472-482, 634-639, 647 (Little Brown and Company, 1984).

[2] *Village of Wilsonville v. S.C.A. Services, Inc.*, 77 Ill. 618, 396 N.E. 2d 552 (1979), *aff'd and remanded*, 86 Ill. 201, 426 N.E. 2d 824 (1981).

[3] Restatement (Second) of Torts, §822.

[4] *Id.*, §821B(2)(a).

[5] Rodgers, *Environmental Law,* 100 (West Publishing Co. 1977).

[6] *Beach v. Sterling & Zinc Co.*, 65 N.J. Eq. 65, 33 A. 286 (1895); *aff'd*, 55 N.J. 824, 38 A. 426 (1897); *Thompson v. Kraft Cheese Co.*, 210 Cal. 171, 291 P. 204 (1931); *Barrington Hills Country Club v. Village of Barrington*, 357 Ill. 11, 191 N.E. 239 (1934); *Rose v. Socony-Vacuum Corp.*, 54 R.I. 411, 173 A. 627 (1934); *Iverson v. Vint*, 243 Iowa 949, 54 N.W. 2d 494 (1952); *Cook Indus. Inc. v. Carlson*, 334 F.Supp. 809 (N.D. Miss. 1971); *Kostyall v. Cass*, 163 Conn. 92, 302 A.2d 122 (1972).

[7] *Boomer v. Atlantic Cement Co.*, 26 N.Y. 2d 219, 257 N.E. 2d 870 (1970). See also *Susquehanna Fertilizer Co. v. Malone*, 73 Md. 268, 20 A. 900 (1890).

[8] 113 Tenn. 331, 83 S.W. 658 (1904).

[9] *Id.* at 666-67.

[10] 161 Cal. 239, 118 P. 928 (1911).

[11] 118 P. at 933.

[12] But see *Village of Wilsonville v. SCA Services, Inc.*, supra, n.2.

[13] *American Smelting & Refining Co. v. Godfrey*, 158 F. 225 (8th Cir. 1907); *DeBlois v. Bowers*, 44 F. 2d 621 (D.C. Mass. 1930); *Bartel v. Ridgefield Lumber Co.*, 131 Wash. 183, 189, 229 P. 306, 308 (1924).

[14] 33 U.S.C. §1311(b)(2)(A).

[15] *Ballard v. Tomlinson*, 29 Ch. D. 115 (1885). See also *Lewis v. Stein*, 16 A. 214 (1849) ("It is a nuisance to throw from day to day into water . . . any substance that renders it less pure and excites disgust in those who use it.")

[16] *Pensacola Gas. Co. v. Pebley*, 25 Fla. 381, 5 So. 593 (1889).

[17] *Id.* at 594-95.

[18] *Id.* at 595.

[19] *Dillon v. Acme Oil Co.*, 2 N.Y. Supp. 289 (1888).

[20] *Id.* at 291.

[21] 12 S.W. 937 (1890). See also *Ball v. Nye*, 99 Mass. 582 (1868).

[22] *Id.* at 938, 939.

[23] Restatement (Second) of Torts, §157; see also *Dandoy v. Oswald Bros. Paving Co.*, 113 Cal. App. 570, 298 P. 1030 (1931).

[24] *Davis v. Georgia Pacific Corp.*, 251 Ore. 239, 445 P.2d 481, 484 (1968).

[25] Prosser, Torts at 69, note 19 (4th ed. 1971).

[26] Wigmore, Responsibility for Tortious Acts, 7 *Harvard L. Rev.* at 315, 383, 441 (1894).

[27] *Id.* at 63-76.

[28] *Phillips v. Sun Oil Co.*, 307 N.Y. 328, 121 N.E. 2d 249 (1954).

[29] *City of Pasadena v. City of Alhambra*, 33 Cal. 2d 908, 207 P.2d 17 (1949).

[30] *Martin v. Reynolds Metals Co.*, 221 Ore. 86, 97, 342 P.2d 790, 794, 795 (1959).

[31] See *Luthringer v. Moore*, 31 Cal. 2d 489, 190 P.2d 1 (1948).

[32] L.R.I. (Ex 265, 270) (1866), *aff'd sub nom, Rylands v. Fletcher*, L.R. 3 (H.L. 330) (1868).

[33] 182 N.J. Super. 210, 219 (App. Div. 1981), *aff'd*, 94 N.J. 254, 463 A.2d 893 (1983).

[34] *Branch v. Western Petroleum, Inc.*, 657 P.2d 267 (Utah 1982).

[35] Restatement (Second) of Torts, §282.

[36] See *Defeo v. People's Gas Co. of New Jersey*, 6 N.J. Misc. 790, 142 A. 756 (1928).

[37] 523 F.Supp. 642, 658 (E.D. Pa. 1981).

[38] 364 So. 2d 604 (La. 1978).

[39] See, for example, *Harper v. Hoffman*, 95 Idaho 933, 523 P. 2d 536 (1974); *Whitt v. Jarnagin*, 91 Idaho 181, 418 P. 2d 278 (1966); *Dumas v. Hartford Accident & Indemnity Co.*, 94 N.H. 484, 56 A. 2d 57 (1947); *Swenson v. Nairn*, 21 N.J. Misc. 70, 30 A. 2d 897 (1943).

[40] See, for example, *United States v. Vertac*, 489 F.Supp. 870 (E.D. Ark. 1980); *United States v. Solvents Recovery Service*, 496 F.Supp. 1127 (D. Conn. 1980).

[41] 42 U.S.C. 9652(d).

[42] See *United States v. Chem-Dyne*, 572 F.Supp. 802 (S.D. Ohio 1985); *United States v. Wade*, 577 F.Supp. 1326 (E.D. Pa. 1983); *Landers v. East Texas Salt Water Disposal Co.*, 248 S.W. 2d 731 (Texas 1982); *Michie v. Great Lakes Steel Div.*, 495 F.2d 213 (6th Cir. 1974).

[43] 129 Cong. Rec. H9153-54 (daily ed., Nov. 3, 1983).

[44] 130 Cong. Rec. 13817 (daily ed., Oct. 5, 1984).

[45] The factual background is drawn from the opinion of the Supreme Court of Rhode Island in *Wood v. Picollo*, 443 A.2d. 1244 (1982).

[46] 54 R.I. 411, 173 A. 627 (1934).

[47] *Id.* at 1249.

[48] *Id.*

[49] The factual background is drawn from the Louisiana Court of Appeals decision in *Ewell v. Petro Processors, et al.*, 364 So.2d. 604 (1972).

[50] *Id.* at 607.

[51] *Id.* at 608.

[52] *Id.* at 609.

[53] *Id.*

Part II

The 1976 RCRA and The 1980 Amendments

Chapter 4

RCRA as Enacted in 1976

President Gerald R. Ford on October 21, 1976, signed into law the Resource Conservation and Recovery Act (RCRA). The enactment of this statute established, for the first time, federal authority to regulate the disposal of hazardous wastes.[1] That authority is spelled out in Subtitle C of RCRA. RCRA does not address, except indirectly, abandoned sites or closed facilities in which hazardous wastes had been handled or disposed of in the past.* The worst of these waste sites are covered by the Comprehensive Environmental Response, Compensation and Liability Act of 1980, commonly referred to as "Superfund."[2] Also, RCRA does not regulate hazardous substances which are within the productive stream of commerce and thus not considered wastes. These chemicals are dealt with under other statutes such as the Toxic Substances Control Act;[3] the Federal Insecticide, Fungicide and Rodenticide Act;[4] and the Hazardous Materials Transportation Act.[5]

RCRA as originally enacted[6] was a relatively straightforward law. Subtitle C of the Act is divided into six major components:

1) Requirements for identifying and listing hazardous wastes;

2) A system for tracking hazardous waste from "cradle to grave";

3) Standards for generators and transporters of hazardous waste and for hazardous waste treatment, storage and disposal (TSD) facilities;

4) A permitting program for TSD facilities;

5) Authorization for states to administer the hazardous waste program; and

6) Provisions concerning enforcement.

Identification of Hazardous Wastes

"Hazardous waste" is defined in RCRA as "a solid waste, or combination of solid wastes, which may, because of its quantity, concentration, or physical, chemical or infectious characteristics . . . pose a substantial present or potential hazard . . . when improperly treated, stored, transported or disposed of, or otherwise mismanaged."[7]

RCRA's use of the adjective "solid" is misleading because "solid waste" is defined in the statute to include liquid, semi-solid, or contained gaseous materials if they are wastes, and also "garbage" and "other discarded material."[8] The definition also states that such wastes must result from industrial, commercial, mining, agricultural operations or community activities, but does not include material from domestic sewage, irrigation return flow, industrial discharges subject to Clean Water Act permitting, or nuclear byproducts.

*It is worth noting that Section 7003 of RCRA, the "imminent and substantial endangerment" provision, authorizes the Department of Justice, on behalf of EPA, to force the cleanup of abandoned or closed hazardous waste sites. The scope of Section 7003, and of its Section 7002 citizen suit counterpart, are discussed below in this chapter and in the chapters dealing with the 1980 and 1984 Amendments to RCRA.

Section 3001, the first section of Subtitle C, required that within 18 months of the statute's enactment—that is, no later than April 21, 1978—EPA establish criteria for identifying hazardous waste. The criteria must include: (1) toxicity, persistence and degradability; (2) potential for accumulation in tissue; and (3) flammability and corrosiveness.[9] Guided by these criteria, EPA was directed to issue regulations which listed particular hazardous wastes and which identified the characteristics of hazardous wastes.

In complying with these requirements, EPA has listed more than 400 specific hazardous wastes. The agency has also established four basic criteria for determining whether a waste is hazardous: ignitability, corrosivity, reactivity, and toxicity.[10]

Tracking System

Section 3002(5) required that by April 21, 1978, EPA establish a manifest system to assure that all hazardous waste that is not disposed of or treated on-site would be delivered to approved treatment, storage or disposal facilities.[11] Regulations issued under this provision specify that generators must designate in the manifest the facility to which the wastes are being transported.[12] There must be enough copies of the manifest for each transporter and the designated facility, with an additional copy to be *returned* to the generator when the waste is ultimately delivered.[13] If the generator does not receive its copy of the manifest within 35 days of the date the waste is accepted by the initial transporter, the generator must investigate and notify the government if there is a problem.[14] In theory, the system works like certified mail: the sender receives written confirmation that the materials sent were in fact received.

Standards for Generators, Transporters and TSD Facilities

Three separate sections, 3002 through 3004, required EPA to issue regulations establishing standards for generators, transporters, and TSD facilities.[15] For generators, these regulations were required, among other things, to establish requirements for recordkeeping, labeling, use of containers, and providing information on the chemical composition of the hazardous wastes. EPA regulations designed to carry out these requirements are published at 40 C.F.R. Section 262. One of the most important duties established by the regulations is the obligation for the generator to determine if any of its solid wastes are hazardous wastes.[16]

For transporters, Section 3003 required that EPA establish requirements for recordkeeping, labeling, and compliance with the manifest system, including transportation only to facilities designated on the form.[17] In addition to implementing these statutory obligations, the regulations, published at 40 C.F.R. Section 263, require transporters to take immediate action to protect human health and the environment in the event of a discharge during shipment.[18]

Section 3004 required that EPA establish performance standards for owners and operators of hazardous waste treatment, storage, and disposal facilities[19] including standards governing operations,[20] location, design, and construction.[21] Further, Section 3004 mandated that the standards include requirements for recordkeeping, monitoring and inspection, contingency plans, personnel training, and financial responsibility. The regulations for TSD facilities on these matters, published at 40 C.F.R. Section 264, also require that the owner-operator monitor groundwater in order to detect and evaluate any migration of contaminants from a facility.[22]

Permit Program

Section 3005 of the 1976 RCRA required that all TSD facilities handling hazardous wastes obtain permits.[23] Through these permits, the federal requirements would be made applicable to individual facilities. Recognizing that EPA would not be able to quickly issue permits for every TSD facility, Congress provided for an "interim status" designation. Those qualifying for interim status, would be treated as having been issued a permit until a final decision was made on their permit application.[24] To qualify for interim status, a facility was required to: (1) have been in existence on the date of enactment of the Act; (2) have complied with the notification requirements of Section 3010(a); and (3) have applied for a permit. The EPA regulations on permit requirements are published at 40 C.F.R. Section 270.

State Authorization

Section 3006 authorized states to assume control of the hazardous waste program from EPA.[25] By April 21, 1978, EPA was to promulgate guidelines to assist in the development of state programs.

Specifically, EPA was required to approve state applications to administer the hazardous waste program unless the state program: 1) was not equivalent to the federal program; (2) was not consistent with the federal program or with state programs applicable in other states; or (3) did not provide adequate enforcement. Despite the equivalence and consistency requirements, RCRA Section 3009 preserves any state's authority to impose requirements which are "more stringent" than the federal requirements.[26]

If at any time the state program failed to comply with the federal guidelines, EPA could withdraw authorization and re-establish a federal program.[27] While that power was granted to EPA, federal withdrawals of state delegations are extremely rare in the environmental field, and there are unlikely to be many cases of such rescissions in the future. EPA regulations establish separate programs for interim and final state authorization. These and other EPA regulations on state authorization are published at 40 C.F.R. Section 256.

Enforcement

Most of EPA's enforcement powers under RCRA were contained in Section 3008.[28] That provision authorized EPA to issue compliance orders assessing fines and setting a compliance schedule 30 days after an offender is notified of its violation. Failure to comply with such an order was made punishable by a civil fine of not more than $25,000 for each day of noncompliance beyond the time specified in the compliance order. EPA or an authorized state was empowered also to revoke or suspend a facility's permit.

Congress also provided that criminal sanctions could be applied to any person who knowingly (1) transported hazardous wastes to a facility that did not have a permit; (2) disposed of hazardous wastes without having obtained a permit; or (3) made a false statement or representation in any application or other document required to be filed. The maximum penalty for a criminal violation was a fine of not more than $25,000 for each day of violation or imprisonment of up to a year, or both. For a second offense, Section 3008 authorized a fine of not more than $50,000 per day of violation, or imprisonment of up to two years, or both.

Perhaps the most important weapon in EPA's enforcement arsenal is provided in Section 7003. Under that provision EPA could bring a suit for injunctive relief upon receipt of evidence that any person's handling, storage, treatment, transportation, or disposal of hazardous waste was presenting "an imminent and substantial endangerment to health or the environment." Because this authority is applicable to solid wastes and not just to wastes formally listed or identified as hazardous, the scope of Section 7003 is extremely broad.

Section 3007 authorized EPA or state employees to inspect the premises and records of any person who generates, stores, treats, transports, disposes, or otherwise handles hazardous wastes.[29] As with the "imminent and substantial endangerment" provision, it was not necessary that the waste be formally identified as hazardous. Section 3007 specifically authorized EPA to copy records and sample wastes. It confined entry to "reasonable times" and required that inspections be commenced and completed with "reasonable promptness." Finally, it required that any information obtained in the inspections be made available to the public. However, the public was not entitled to obtain trade secrets and other confidential information protected by 18 U.S.C. Section 1905.

In addition to the enforcement tools provided to the government, the 1976 Act also authorized citizens to sue the Administrator of EPA for failure to perform a nondiscretionary duty under the Act. Section 7002 also allowed citizens to sue any person (including a state or the federal government) for violation of any RCRA permit or regulation.[30] For both types of actions, plaintiffs were required to give 60 days notice of their intention to sue. However, no notice was required for violations of Subtitle C. Courts were authorized to award litigation costs including reasonable attorney and expert witness fees to any party whenever the court determined such award was appropriate.

Chapter 4 Footnotes

[1] 42 U.S.C. §§6901 *et seq.*
[2] 42 U.S.C. §§9601 *et seq.*
[3] 15 U.S.C. §§2601 *et seq.*
[4] 47 U.S.C. §§135 *et seq.*
[5] 42 U.S.C. §§1801 *et seq.*
[6] 90 Stat. 2795.
[7] RCRA Sec. 1004(5), 42 U.S.C. §6903(5).
[8] RCRA Sec. 1004(27), 42 U.S.C. §6903(27).
[9] RCRA Sec. 3001(a), 42 U.S.C. §6921(a).
[10] 40 C.F.R. §261, Subpart D.
[11] 42 U.S.C. §6922(5).
[12] 40 C.F.R. §262.20.
[13] 40 C.F.R. §*262.22.*
[14] 40 C.F.R. §262.42.
[15] 42 U.S.C. §§6922-24.
[16] 40 C.F.R. §262.11.
[17] 42 U.S.C. §6923.
[18] 40 C.F.R. §263.30(a).
[19] 42 U.S.C. §6924.
[20] 42 U.S.C. §6924(3) and (6).
[21] 42 U.S.C. §6924(4).
[22] 40 C.F.R. §264.90.
[23] 42 U.S.C. §6925.
[24] RCRA Sec. 3005(e), 42 U.S.C. §6925(e).
[25] 42 U.S.C. §6926.
[26] 42 U.S.C. §6929.
[27] RCRA Sec. 3006(e), 42 U.S.C. §6926(e).
[28] 42 U.S.C. §6928.
[29] 42 U.S.C. §6927.
[30] 42 U.S.C. §6972.

Chapter 5

1980 Amendments

In 1980 Congress passed two statutes that amended RCRA.* The Used Oil Recycling Act[1] added requirements involving recycled oil. Section 7 of that Act, adding Section 3012 to RCRA,[2] required that by October 15, 1981, EPA promulgate standards necessary to protect public health and the environment from hazards associated with recycled oil. It specified, however, that the regulations must "not discourage the recovery or recycling of used oil."[3]

More extensive changes and additions were made by the Solid Waste Disposal Act Amendments of 1980,[4] although that law did not embark on new policy directions. The principal changes made by the 1980 Amendments were to (1) authorize EPA to regulate new facilities more stringently than existing ones; (2) sanction explicitly the concept of "interim status standards" that had been devised by EPA, and extend the time to qualify for interim status; (3) increase penalties for criminal violations; (4) provide various additional authorities to assist EPA in carrying out its duties; and (5) suspend regulation of certain wastes while EPA studied them to determine if they should be listed as hazardous.

Interim Status Standards

The 1980 Amendments gave congressional sanction to the role EPA administratively had devised for interim status in the regulatory process. Nothing in the original statute suggested that Congress contemplated "interim status standards," as opposed to implementing final standards in permits. EPA nonetheless imposed them, under a somewhat strained reading of Section 3004(7).[5] Any argument that EPA exceeded its authority by establishing "interim status standards" was deliberately resolved by Section 13 of the 1980 Amendments, which provided criminal penalties for certain violations of "applicable interim status regulations and standards."[6]

Congress also expanded the scope of interim status by extending eligibility from facilities in existence on October 21, 1976, to those in existence on November 19, 1980.[7] According to the Conference Report: "This extension of time is necessitated both by the delay in promulgation of applicable regulations and the desire of the conferees to encourage the operation of new, technologically modern treatment and disposal facilities."[8] The Report also stated that to be eligible a facility need not actually be in operation by November 19, 1980, but rather it must have obtained all necessary state and local permits and have made financial commitments that cannot be terminated.[9]

Authority to Regulate New Facilities More Stringently than Existing Ones

Section 9 authorized EPA to establish more stringent standards for new facilities than for existing ones.[10] The basis of this change is that modifying

*The Quiet Communities Act of 1978, 42 U.S.C. §§4901-4918, made several changes of a technical nature to the Act.

existing facilities is frequently more costly and technically difficult than installing the desired feature in original construction.[11]

Public Participation in New Facility Permitting

Section 26 required that, before issuing a permit for a new facility, EPA must give notice in local newspapers and over local radio stations and also provide written notice to units of local government having jurisdiction. If EPA received written notice of opposition to issuance of the permit, it was required to hold an informal public hearing.[12]

Changes in Enforcement Authority and Penalties

Section 13 made a number of important changes to RCRA's[13] enforcement provisions. It deleted the requirement that EPA give notice and then wait 30 days before issuing a compliance order.[14] Also, it allowed EPA to initiate a civil action directly rather than first issuing a compliance order.[15]

The maximum criminal fine was doubled to $50,000 per day for violations related to hazardous waste management activities which were undertaken either without a permit or in violation of a permit. Section 13 also clarified that the criminal provisions covered knowing violation of a material permit condition.[16] It also made clear that only "material" false statements are subject to criminal sanctions.[17] In addition, a new subsection, 3008(d)(4), provided criminal penalties for knowingly destroying, altering, or concealing any records required to be maintained.

Perhaps the most significant change made by the 1980 Amendments to the enforcement provisions involved the creation of a new criminal offense of "knowing endangerment." Specifically, new subsection (e) made it a crime for any person to *knowingly* place another individual in danger of death or serious bodily injury by transporting hazardous waste to, or treating, storing, or disposing of it at a facility without a permit or interim status; failing to comply with interim status standards; or omitting material information from a permit application. If the maximum penalty were applied, a person convicted of "knowing endangerment" could be fined up to $250,000 and imprisoned for five years; an organization was subject to a fine of up to $1 million.

The 1980 Amendments also provided extensive guidance on the necessary state of mind for a criminal offense under the knowing endangerment provision and specified certain defenses.*[18] It is worth noting that prosecution of the knowing endangerment offense did not depend on proof of actual harm. Like reckless driving, it was the creation of the danger that constituted criminal conduct. The requirements regarding the state of mind standard were drawn from the Senate Judiciary Committee's proposed Criminal Code Revision of

*For example, Section 3008(f)(3) established an affirmative defense if "the danger and conduct charged were reasonably foreseeable hazards of an occupational, a business or a profession . . . "

1980, and the Conference Report noted that these requirements differed from the traditional standards in criminal law.[19] Given the complexity of the provision, it is not surprising that only one indictment alleging knowing endangerment was ever handed down.

Finally, Section 12 added as a criminal offense the knowing and willful disclosure of trade secret and other confidental information protected by 18 U.S.C. Section 1905.[20]

Imminent Hazard Administrative Orders

Section 25 expanded the scope of Section 7003 by authorizing EPA to exercise its power to abate imminent hazards through the issuance of administrative orders, as well as by instituting litigation.[21] The scope of permissible administrative orders was defined broadly to include monitoring or testing as well as restraining operations and instituting remedial measures. Congress made violation of an administrative order issued under the imminent and substantial hazard provision subject to a fine of $5,000 for each day of noncompliance.

Testing and Monitoring Requirement

Section 17 authorized EPA to require the present or former owner or operator of an active or inactive waste disposal site to conduct testing, analysis, and monitoring to determine the nature and extent of the danger where EPA determines that there is a substantial hazard to human health or the environment.[22] Failure to comply with an order issued under this section can result in a civil penalty of up to $5,000 for each day of noncompliance.[23]

Private Contractor for Inspections

Section 12 allowed EPA to use private contractors to inspect facilities.[24] This provision expands the agency's enforcement arm beyond its own (usually severely limited) staff resources in conducting inspections of hazardous waste facilities. The Conference Report states, however, that this expanded authority must be exercised judiciously; EPA should avoid the use of contractors who may have a conflict of interest.[25]

Judicial Review

Section 27 made a number of changes in the judicial review provision.[26] Previously, judicial review had been subject to normal federal district court procedures except that review of a regulation or requirement promulgated under the Act was subject to review exclusively in the U.S. Court of Appeals for the District of Columbia. The 1980 Amendments made special provision for review of EPA's action on a permit application or interim authorization in the Circuit Court of Appeals in which the petitioner resides or transacts business. Petitions for review in all the Courts of Appeals were required to be filed within 90 days of the agency action being challenged.

Temporary Exemption for Certain Petroleum, Utility, Mining and Cement Wastes

Section 7 temporarily exempted from hazardous waste regulation four types of wastes.[27] These wastes were 1) certain drilling muds, associated with petroleum operations; 2) certain utility industry wastes; 3) certain mining wastes; and 4) cement kiln wastes.

In each case, EPA was required to conduct a detailed study of the hazards, if any, posed by these wastes. Pending the completion of the EPA studies (following its study of certain mining wastes, EPA announced in June 1986 that it intended to regulate mining wastes under Subtitle D), these wastes could not be subject to regulation under Subtitle C. They could, however, be regulated by other federal law (including Subtitle D) and by states. Additionally, Section 7 of the 1980 Amendments prevented drilling muds from being regulated as "hazardous waste" unless Congress specifically authorized such regulation after reviewing the results of EPA's study.[28]

Interior Department's Exclusive Authority over Certain Mining Wastes

Section 11 conferred on the Secretary of the Interior, rather than on the Administrator of EPA, the exclusive responsibility for carrying out any requirement of Subtitle C with respect to coal mining wastes or overburden for which a surface coal mining and reclamation permit has been issued under the Surface Mining Control and Reclamation Act of 1977.[29]

Generator's Reponsibilities to Ensure that Wastes Arrive at TSD Facilities

Section 8 required EPA to promulgate regulations mandating use of the manifest system to assure that hazardous waste actually arrives at the permitted facility designated on the manifest. EPA was authorized to use other means in addition to the manifest system to accomplish this purpose.[30]

State Inventories of Hazardous Waste Sites

Section 17 required each state to undertake a continuing program to compile, publish, and submit to EPA an inventory of the hazardous waste sites in the state, giving their location; description of the hazards; the name, address, and corporate headquarters of the owner of the site; and other information.[31] Much of the information compiled under the inventory program has been used by states in proposing candidates for Superfund's National Priority List of hazardous waste sites.

State Authority to Impose More Stringent Standards

The 1976 RCRA had implicitly allowed states to impose more stringent standards than the federal government by providing that states and political subdivisions could not impose "less stringent" requirements. Section 4 of the 1980 Amendments made this authority explicit by providing: "Nothing in this title shall be construed to prohibit any state or political subdivision thereof from imposing any requirements, including those for site selection, which are more stringent than those imposed by such regulations."[32]

Chapter 5 Footnotes

[1] P.L. 96-463, 94 Stat. 2055.
[2] Now Section 3014, 42 U.S.C. §6935.
[3] 42 U.S.C. §6935(a).
[4] P.L. 96-482, 94 Stat. 2334.
[5] 42 U.S.C. §6924. See 45 Fed. Reg. 33158-59 (May 19, 1980).
[6] Amending RCRA Sec. 3008(e)(1)(b)(ii), now RCRA Sec. 3008(d)(2)(C), 42 U.S.C. §6928(d)(2)(C).
[7] Amending RCRA Sec. 3005(e), 42 U.S.C. §6925(e).
[8] House Conference Report No.1444, 96 Cong. , 2nd Sess. 33, 34, reprinted in 1980 U.S. Code Cong. & Ad. News 5028, 5033, 5034.
[9] House Conference Report at 34, 1980 U.S. Code Cong. & Ad. News 5028, 5034.
[10] Amending RCRA Sec. 3004(a), 42 U.S.C. §6924(a).
[11] House Conference Report at 34, 1980 U.S. Code Cong. & Ad. News 5028, 5033.
[12] Amending RCRA Sec. 7004(b), 42 U.S.C. §6974(b).
[13] 42 U.S.C. §6928.
[14] Amending RCRA Sec. 3008(a), 42 U.S.C. §6928(a).
[15] Id.
[16] Amending RCRA Sec. 3008(d)(2), 42 U.S.C. §6928(d)(2).
[17] Amending RCRA Sec. 3008(d)(3), 42 U.S.C. §6928(d)(3).
[18] Adding RCRA Sec. 3008(f), 42 U.S.C. §6928(f).
[19] House Conference Report at 39 and 40, 1980 U.S.Code Cong. & Ad. News 5028, 5038, 5039.
[20] Adding RCRA Sec. 3007(b)(2), 42 U.S.C. §6927(b)(2).
[21] Amending RCRA Sec. 7003, 42 U.S.C. §6973.
[22] Adding new provision RCRA Sec. 3013, 42 U.S.C. §6934.
[23] Adding RCRA Sec. 3013(e), 42 U.S.C. §6934(e).
[24] Amending Sec. 3007(a), 42 U.S.C. §6927(a).
[25] House Conference Report at 35, 1980 U.S. Code Cong. & Ad. News 5028, 5035.
[26] Amending RCRA Sec. 7006, 42 U.S.C. §6976.
[27] Adding RCRA Sec. 3001(b)(2) and (3), 42 U.S.C. §6921(b)(2) and (3).
[28] Adding RCRA Sec. 3001(b)(2), 42 U.S.C. §6921(b)(2).
[29] Adding RCRA Sec. 3005(f), 42 U.S.C. §6925(f).
[30] Amending RCRA Sec. 3002(5), 42 U.S.C. §6922(5).
[31] Adding RCRA Sec. 3012, 42 U.S.C. §6933.
[32] Amending RCRA Sec. 3009.

Part III

The 1984 Amendments

Chapter 6

Land Disposal Restrictions and the "Hammer" Provisions

Summary

Land Disposal Bans:

• After November 8, 1984, no hazardous wastes may be disposed of in salt domes, salt bed formations, mines, or caves unless EPA determines that such disposal is protective of human health and the environment.

• After May 8, 1985, no bulk or noncontainerized liquid hazardous wastes may be disposed of in a landfill.

• After November 8, 1985, the disposal of nonhazardous liquids in a hazardous waste landfill is prohibited unless EPA determines that there are no other reasonable disposal alternatives and the disposal will not risk contaminating underground drinking water supplies.

• After November 8, 1986, specific dioxins and solvents may not be land disposed (except by underground injection) unless EPA determines the prohibition is not required to protect human health and the environment for as long as the waste remains hazardous.

• After July 8, 1987, no California list wastes may be land disposed (except by underground injection) unless EPA, using the standard above, determines the prohibition is not necessary.

• After August 8, 1988, no California list waste or specified dioxin and solvent waste may be disposed of by underground injection unless EPA, using the standard above, determines the prohibition is not necessary.

• After August 8, 1988, one-third of all remaining hazardous wastes specified by EPA are barred from land disposal facilities, except the ban is extended where the facility meets the new minimum technological requirements and land disposal is the only practicable alternative currently available.

• After June 8, 1989, another third of the remaining hazardous wastes specified by EPA are barred from land disposal facilities unless the facility meets the conditions for extending the deadline for prohibition.

• After May 8, 1990, all listed and identified hazardous wastes are barred from land disposal unless EPA determines that land disposal of particular wastes will be protective of human health and the environment for as long as the wastes remain hazardous.

Extensions of Deadlines for Land Disposal Bans on Dioxins, Solvents, California Wastes, and EPA Listed Remaining Wastes:

• EPA may grant a case-by-case variance of up to one year where applicant shows that there is a binding contract for alternative disposal which for circumstances beyond its control is not available.

• EPA may grant on a waste-by-waste basis a variance for up to two years for lack of adequate alternative treatment capacity for a waste.

Other Important Provisions:

• Where any hazardous waste is prohibited from land disposal, storage is also prohibited unless solely for purpose of accumulating quantities sufficient for proper disposal.

• Waste or used oil contaminated with dioxin or any other RCRA-listed or identified waste is banned from use in dust suppression or road treatment.

• EPA must promulgate regulations 1) minimizing disposal of containerized liquid hazardous waste in landfills; 2) establishing treatment standards which will substantially diminish the toxicity of waste; and 3) establishing requirements governing air emissions at hazardous waste facilities.

If the Administrator fails to promulgate regulations, or make

> a determination . . . for any hazardous waste within 66 months after the date of enactment of the Hazardous and Solid Waste Amendments of 1984, such hazardous waste shall be prohibited from land disposal.[1]

This one sentence is the single most significant of a number of "hammer" provisions in the 1984 RCRA Amendments. It establishes a self-enforcing deadline for ending America's reliance on land disposal of hazardous waste. The first appearance of a "hammer" provision in environmental legislation occurred in 1982 when Congressman James J. Florio of New Jersey introduced H.R. 6307, the RCRA reauthorization bill of the 97th Congress. It was included in the section on small quantity generators and provided that if EPA failed to promulgate regulations governing wastes produced by small quantity generators by a specific deadline, the regulations designed for "large" quantity generators automatically would be imposed on "the little guys." "It would hit them like a hammer," protested Joe Gerard of the American Furniture Manufacturers Association. While the phrase achieved instant popularity, the concept was hotly debated.

The problem, as virtually everyone agreed, was that EPA wasn't moving fast enough in regulating the management of hazardous wastes. The agency's failure to meet the deadlines established in the original Act resulted in extensive litigation[2] and massive frustration. According to its proponents, the "hammer" would provide the tool for smashing the Gordian Knot at EPA and for getting results. Result-oriented Members of Congress were drawn to the idea that, if EPA would not do the job, Congress would do it. After all, they reasoned, it was Congress' basic responsibility to write the rules in the first place. Therefore, it came as no surprise that on April 11, 1983, during the House Commerce, Transportation and Tourism Subcommittee consideration on H.R. 2867, a hammer provision was added to Section 5, the provision on land disposal of hazardous waste.

Although the hammer amendment to Section 5, offered by Congressman Dennis Eckart (D-Ohio), was adopted with little discussion, the hammer strategy itself had not suddenly become an acceptable method of legislating. On the contrary, the "Minority Views" to the House Report on H.R. 2867, offered by 10 Republicans on the Energy and Commerce Committee, made clear that the hammer wasn't their idea of sensible public policy.

> . . . H.R. 2867 requires the Administrator to determine whether or not hazardous waste should be prohibited from one or more methods of land disposal . . . If no determination is made, these wastes are also automatically prohibited from all methods of land disposal. These provisions actually force EPA to make a negative finding, that is, a determination that a land disposal prohibition is not needed to protect health and the environment, or a ban

> automatically takes effect. This idea of regulating or banning something by operation of law as a means to force an administrative decision is unprecedented.
>
> We in Congress do not have the technical expertise to declare such prohibitions merely because EPA has missed an arbitrary deadline. How do we know that the specific wastes [that are] listed . . . should be prohibited from all forms of land disposal? How do we know that the concentration levels specified are appropriate? These are regulatory decisions which should not be made by Congress.
>
> This approach of automatic regulation absent a negative finding is being used to exert pressure on EPA to act within legislative deadlines. The rationale is that the affected community will lobby EPA to act in time, for fear of facing automatic prohibitions. Why should we draft laws which penalize the regulated community because EPA does not act in a timely fashion, or Congress has set unrealistic deadlines? We feel that this approach is technically and intellectually inappropriate.[3]

Despite this admonition by Ranking Minority Member James T. Broyhill (R-North Carolina) and most of the other Republicans on the Energy and Commerce Committee, the hammer provision, although modified, was retained in the final version of the legislation.

In addition, several other self-implementing requirements and prohibitions were included in the land disposal provisions. This approach was combined with unusually detailed instructions to EPA on how to regulate virtually every aspect of the disposal system—from the allowable permeability of liners for surface impoundments to the concentrations at which certain chemical wastes must be banned from land disposal.

At the close of the reauthorization process, the need for radical legislative change, including the hammer provisions, was explained by Senator George Mitchell of Maine:

> It has become evident that a strong congressional expression of disapproval of EPA's slow and timid implementation of the existing law is necessary, as well as a clear congressional direction mandating certain bold, preventive actions by EPA which will not be taken otherwise, despite the existing, broad authorities contained in RCRA. EPA has not implemented the Resource Conservation and Recovery Act aggressively. The Agency has missed deadlines, proposed inadequate regulations, and even exacerbated the hazardous waste problem by suspending certain regulations.
>
> It has become evident that this slow, plodding course will be continued in the absence of a clear congressional directive. This is not acceptable.[4]

The clear congressional directive to change course and increase speed is abundantly demonstrated in Section 201,[5] governing "Land Disposal of Hazardous Waste."

The goal of this provision, as set forth in Section 1001(b), is simple and straightforward: "[T]o avoid substantial risk to human health and the environment, reliance on land disposal should be minimized or eliminated."[6] To accomplish this goal, Congress employed both the hammer and its first cousin, the flat-out prohibition, in a complicated series of limitations on specific methods and locations of land disposal. First, in an attempt to prevent any recurrences of the Times Beach phenomenon, Congress banned outright the use of waste oil or other material which is contaminated or mixed with dioxin or any other hazardous waste for the purpose of dust suppression or road treatment. Another ban is placed on the disposal of hazardous waste into or above a "formation" which contains an underground source of drinking water that is within one-quarter mile of the injection well.[7] Also, effective May 8, 1985, the placement in any landfill of bulk or noncontainerized liquid hazardous waste—whether or not absorbants have been added—is prohibited.[8] In its Codification Rule*, the agency has interpreted this ban to prohibit even storage of material awaiting further treatment or disposal, and to preclude use of such storage locations as treatment chambers.[9] Additionally, starting November 8, 1985, the Amendments prohibit the placement of *any* nonhazardous liquid in a Subtitle C landfill except in certain very limited circumstances.[10]

In a prohibition on the location of certain storage and disposal practices, Congress prohibited "the placement of any hazardous waste in a salt dome formation, salt bed formation, underground mine, or cave" unless EPA (1) determines that such placement will be protective of human health and the environment, (2) promulgates performance standards for a permit, and (3) issues a permit. The other land disposal restrictions are "hammers," rather than absolute prohibitions, with bans taking effect if EPA does not make a specified determination within the prescribed time period. Under the first hammer provision, effective November 8, 1986, certain dioxin-containing wastes and certain waste solvents will be prohibited from all forms of land disposal (except deep well injection)[11] unless the Administrator determines that the prohibition is not required in order to protect human health and the environment.

Eight months later, on July 8, 1987, disposal of the so-called "California list" wastes** will be prohibited from all forms of land disposal (except

*The Codification Rule issued by EPA on July 15, 1985 (50 Fed. Reg. 28702), amends existing hazardous waste regulations to reflect those statutory provisions of the 1984 Amendments that become effective as a matter of statute in the short term. In the Codification Rule, EPA also made a number of interpretations of the Amendments, the most important of which are presented throughout this book in the pertinent sections.

**"California list" wastes are those which the State of California targeted for prohibition from land disposal in legislation enacted in 1982 in Section 66900 of the California Health Code. The wastes, set forth in full in the 1984 Amendments (Section 201(d)(2), adding RCRA Section 3004(d)(2)), include liquid wastes and sludges containing set concentrations of specified

underground injection) unless the EPA Administrator, following the same standard as used for solvents and dioxins, determines that such disposal is safe. EPA must also decide whether to prohibit underground injection of the California list wastes and of solvents and dioxins by August 8, 1988.

All remaining hazardous wastes, both listed and identified are to be evaluated over a five-and-a-half year period ending May 8, 1990.[12] The criteria for evaluating these remaining wastes is somewhat more stringent than for the dioxins, solvents and "California list" wastes. To permit continued land disposal of the remaining listed and identified wastes, the Administrator must find that such disposal "*will be protective* of human health and the environment for as long as the waste remains hazardous." [Emphasis added.]

In making determinations concerning the safety of land disposal (including underground injection), the Administrator must take into account:

1) the long-term uncertainties associated with hazardous waste disposal;

2) the goal of managing hazardous wastes in an appropriate manner in the first instance; and

3) the persistence, toxicity, and mobility of hazardous wastes and their propensity to bioaccumulate.*

In explaining the standard of review required by Section 201, the Conference Report, echoing the statutory language, states that:

> A method of land disposal may not be determined to be protective of human health and the environment if a specified waste contains significant concentrations of one or more hazardous constituents that are highly toxic, highly mobile or have a strong propensity to bioaccumulate, unless it is demonstrated to a reasonable degree of certainty that there will be no migration of such constituents from the disposal unit or injection zone for as long as the waste remains hazardous."[13]

The Act directs the Administrator to conduct a waste-by-waste review of all the remaining wastes in accordance with a schedule that the Administrator must develop by November 8, 1986.[14] The schedule must set forth a plan by which EPA will make decisions regarding land disposal for one-third of all listed wastes by August 8, 1988, two-thirds of all listed wastes by June 8, 1989, and all remaining listed and identified wastes by May 8, 1990. EPA also is required to make land disposal prohibition decisions for all wastes listed or identified after November 8, 1984, within six months of the identification or listing.[15] The Senate and House conferees gave considerable thought to the criteria for selecting which wastes should be evaluated first. As a general rule, they decided that the high volume wastes that are intrinsically the most hazar-

heavy metals and arsenic, highly acidic liquids, organic liquids containing 50 or more ppm PCBs, and halogenated compounds in concentrations of 1,000 mg/kg or more.

*These same three factors must be taken into account by the Administrator in making the decisions whether to prohibit land disposal of the dioxins, solvents, and California list wastes.

Table 6

New EPA Activities Mandated By 1984 RCRA Amendments[a]

Activity	Statutory Deadline
Expand Regulated Waste Coverage	
Small quantity generator study	4 months
Small quantity generator regulations	17 months
New listings of hazardous substances	6 months
New toxicity testing procedures	28 months
New characteristics of hazardous wastes based on organic toxicity	24 months
Change Waste Management Practices	
Ban on land disposal of bulk liquids in landfills	6 months
Ban on land disposal of high priority hazardous wastes	32 months
Ban on land disposal of solvents and dioxins	24 months
Minimum technological requirements for land disposal facilities	Immediate
Potential bans for first one-third of EPA's listed wastes	45 months
Potential bans for second one-third of EPA's listed wastes	55 months
Potential bans for third one-third of EPA's listed wastes	66 months
Ban on injecting wastes above or into drinking water aquifer	6 months
Standards for acceptable treatment technologies to diminish toxicity or risk of exposure	Concurrent with prohibitions
Regulate Additional Activities	
Interim construction standards for underground storage tanks	4 months
Performance regulations for existing and new underground storage tanks	30 months
Salt dome storage performance standards	No limit

dous should be scheduled for early review. However, the Conference Committee made clear that EPA was not expected to undertake a rigorous assessment of risk for the purpose of developing a schedule. Moreover, while volume and toxicity are two critical factors, those two considerations

> are not the only relevant ones and in some cases may not give an accurate indication of which wastes present the greatest risks. For example, if a highly toxic waste degrades very rapidly, it may not present a serious risk in the land disposal environment. Likewise, a particular waste may not be very toxic relative to other hazardous wastes but may be so mobile (or in fact may mobilize other relatively immobile wastes) that it does in fact present a serious risk in land disposal. In establishing the schedule, EPA may consider factors other than toxicity and volume if they are relevant to the risk of the waste in land disposal.[16]

In addition to establishing the criteria for EPA to use in devising the schedule, Congress wanted to make sure that this rulemaking would not be subject to normal delays. Thus, it exempted EPA's development of the schedule

Table 6

New EPA Activities Mandated By 1984 RCRA Amendments[a]

Activity	Statutory Deadline
Regulations to minimize land disposal of hazardous liquids	16 months
Regulations for deep well injection of high priority wastes, dioxins, and solvents	45 months
Standards for monitoring and control of air emissions from treatment, storage, or disposal (TSD) facilities	30 months
Standards for leak detection systems for land facilities	30 months
Standards for areas of vulnerable hydrology precluding siting of TSD facilities	18 months
Regulations on blending and burning hazardous wastes	24 months
Regulations on recordkeeping for blending and burning	15 months
Regulations on transporting fuels with hazardous wastes	24 months
Final permits for all TSD facilities	48 months
Final Incinerator Permits	60 months
Standards on generation and transportation of used oil for recycle	24 months
Regulations on exporting hazardous wastes from the United States	12 months
Ruling on the hazardousness of discharges from publicly owned sewage treatment works	18 months

SOURCE: Congressional Budget Office, Hazardous Waste Management: Recent Changes and Policy Alternatives" (May 1985). Based on the 1984 RCRA amendments.

a. Does not include several studies and inventories that the EPA must also perform within 36 months of enactment and several notification and certification activities required of private industry.

from the Paperwork Reduction Act. Preparation of the schedule is also exempted from formal rulemaking, i.e., hearings on the record.[17] Further, to assure that the schedule would not be tied up in protracted litigation, Congress exempted it from judicial review. In so doing, the conferees made clear that, while the substance of the schedule, which is left entirely to agency discretion, may not be challenged, the requirement to develop and publish the schedule in a timely manner is a mandatory duty that is subject to judicial enforcement.[18]

After considerable debate on the utility of hammers, the conferees finally decided that if EPA fails to make a timely decision on a waste that has been placed in either the 45- or 55-month group, that waste can be disposed of in a landfill or surface impoundment *only* if the facility meets all applicable minimum technology requirements, including double-liners and leachate collection systems. Furthermore, prior to disposal, the generator of the waste must certify that the availability of alternative treatment capacity has been investigated, and that use of the landfill or surface impoundment is the only practical means of disposal. If by May 8, 1990—66 months after enactment—EPA has failed to decide the land disposal fate of a particular hazardous waste, that waste is prohibited from all forms of land disposal.[19] This combination of hard and soft hammers represents a compromise forged in the Conference Committee, between those who wanted certainty that most hazardous wastes eventually would be banned from land disposal and those who doubted that alternative technologies would be available by the end of the decade. Also, it separates the decision about the safety of land disposing a particular waste from the decision about whether there is an alternative capacity for that waste. Further, it provides the treatment industry advance notice of a coming market for which they can develop capacity.

To modify some of the rigidity of the hammer, Section 201 allows the Administrator to grant certain limited variances, on a waste-by-waste basis, from the land disposal prohibitions. These variances, really extensions of the deadlines, can be granted for no more than two years after the original effective date *if* the Administrator determines that there is no adequate alternative treatment recovery or disposal capacity.[20] The extension applies to a particular waste at land disposal facilities generally, not to individual petitioning facilities. Additional extensions of up to one year, renewable once, may be granted on a case-by-case basis where an applicant demonstrates that it has a "binding contractual commitment to construct or otherwise provide alternative disposal capacity" which cannot be made available by the deadline because of circumstances beyond the control of the applicant.[21] While these extensions are in effect, the wastes must go to landfills or surface impoundments that meet the minimum technologies and groundwater monitoring requirements.

A categorical exemption is available under the statute for any disposal of contaminated soil or debris resulting from a response action taken under Section 104 or Section 106 of Superfund, or from a corrective action under Subtitle C of RCRA.[22] This exemption, which expires on November 8, 1988, is intended to ensure that Superfund clean-up efforts are not impeded by the lack of alternative treatment capacity.

Section 201 also proscribes storage of a hazardous waste prohibited from land disposal, unless it is stored solely for the purpose of accumulation prior to proper recovery, treatment, or disposal.[23]

Although the statutory bias against land disposal appears inflexible, the statute leaves open the possibility of waste stream modification so that wastes would not be subject to the land disposal prohibitions in the first place. Specifically, EPA is directed to promulgate regulations specifying those levels or methods of treatment which substantially diminish the toxicity of the waste or substantially reduce the likelihood of migration of the hazardous constituents from the waste.[24] These regulations, which are to be promulgated simultaneously with the land disposal restrictions, are intended to ensure that both short-term and the long-term threats to human health and the environment from treated wastes are "minimized."[25] Once a waste is properly treated in accordance with EPA specifications, the land disposal prohibitions would not apply. However, the treated waste that escapes the land disposal prohibition must be placed in a facility that meets all Subtitle C requirements. Depend-

Table 7

Range of Estimated Annual Incremental Costs To Industry Of 1984 RCRA Amendments, By 1990 (In millions of 1983 dollars)

Program Element	Annual Cost
Land Disposal Prohibition	2,650–5,422
Sanitary Landfill (Subtitle D)[a]	1,000–2,000
Burning and Blending Requirement[b]	456–1,620
Small Generators[b]	100– 300
New Technological Requirements[b]	40– 70
Total	4,246–9,417

SOURCE: Congressional Budget Office, "Hazardous Waste Management: Recent Changes and Policy Alternatives" (May 1985). In part based on data obtained from the Environmental Protection Agency.

a. The Subtitle D program includes retrofit requirements for sanitary landfills, such as municipal solid waste landfills. Program costs are uncertain, because it is difficult to predict how many facilities will be required to meet the more stringent standards applicable to hazardous waste landfills. The estimate in this table, therefore, does not include any of the corrective action requirements of new RCRA Section 3004(u). Assuming all Subtitle D facilities must install groundwater monitoring systems, corrective action (40 CFR 264.100) is required at only 20 percent of municipal facilities and 10 percent of industrial facilities, and liners are required at 20 percent of municipal sites and 5 percent of industrial facilities, annual costs could increase by $4 billion to $7 billion. Full application of current hazardous waste landfill standards to all Subtitle D facilities could increase annual costs by $10 billion to $25 billion.

b. Estimated by the Environmental Protection Agency, Office of Policy Analysis (1984).

ing on the flexibility provided in the EPA regulations, it is possible that there could be a substantial amount of hazardous waste treatment aimed specifically at avoiding the land disposal restrictions.

There can be little doubt, however, that Section 201, if implemented aggressively, will prove to be the most costly provision of the 1984 Amendments. The Congressional Budget Office has estimated that the annual additional cost to industry will be between $2.5 billion and $5.5 billion (see chart). Yet, while the 1984 Amendments' policy on restricting land disposal is a bold and expensive experiment, it is one that many observers felt was long overdue. Nevertheless, as demonstrated in the following section, the approach proposed by EPA in January 1986 was not the type of experiment that several conferees had in mind when they wrote the law.

Early Controversy—EPA's Proposed Land Disposal Rules

"The most significant rulemakings required by the Hazardous and Solid Waste Amendments of 1984." That is the description EPA Assistant Administrator J. Winston Porter used in describing the agency's proposed land disposal regulations in February 1986. Porter used the term in testifying before the House Energy and Commerce Committee's Subcommittee on Commerce, Transportation and Tourism just about a month after the 164-page proposal was published in the Federal Register.[26]

The unusual thing about Porter's testimony was that it was one of two he was giving before House and Senate committees on the agency's proposal under Section 201 of the 1984 Amendments. It is not unusual for EPA officials to testify before a congressional subcommittee on a new regulation, but to be called before both subcommittees of both houses of Congress for a regulation still in the proposal state usually means trouble. This case was no exception. The focus of the hearings was EPA's interpretation of Section 201(5)(g) of the 1984 Amendments, in which Congress specified that "a method of land disposal may not be determined to be protective of human health and the environment . . . unless it has been demonstrated to the Administrator, to a reasonable degree of certainty, that there will be no migration of hazardous constituents from the disposal unit or injection zone for as long as the wastes remain hazardous."

EPA in its January 14, 1986, proposal in effect was taking an "environmental results" risk assessment approach to determining which wastes should be treated prior to being disposed of in land. In its proposal, EPA said it was

> . . . not reading the statutory definition of "protective" as an absolute no migration standard. The Agency finds significance in Congress' inclusion of the clause "for as long as the wastes remain hazardous" as a modifier to what otherwise would have been a strict no migration standard, i.e., "no migration of hazardous constituents from the disposal unit." The Agency is today proposing to interpret inclusion of this modifying clause as implying that the statutory standard allows for some migration of hazardous constituents beyond the immediate confines of the disposal unit, as long as such migration does not present any threat to human health and the environment. Support for this interpretation is provided in the legislative history, where Congress noted: "[t]he Administrator is required to find that the nature of the facility and the waste will assure that migration of the wastes will not occur while the wastes still retain their hazardous characteristics in such a way that would present any threat to human health and the environment" (S.Rep.No. 98-284, 98th Cong., 1st Sess. 15 (1983)). Accordingly, EPA is proposing to develop the screening levels identifying "protective" levels [of hazardous constituents] in land disposal . . .[27]

Using this risk analysis approach, EPA proposed interpreting the "no migration" language to allow continued land disposal of hazardous wastes so long as hazardous constituents were not detected 500 feet from the disposal unit in concentrations deemed harmful to human health or the environment. Congress on the other hand, in passing the 1984 Amendments, felt it was taking something more akin to the "best available technology" approach it had taken in certain other pollution control laws under which across-the-board treatment of hazardous wastes was virtually a given.

Despite EPA's development of an intricate methodology for determining "screening levels" (and the agency's selective use of legislative history) several Members of Congress contended that the 1984 Amendments did not authorize the deployment of a risk analysis model to implement Section 201. According to Congressman Dennis Eckart, "the effort made by the agency to mangle the plain language of the statute is nothing short of astonishing . . . ."

> To construe these words in a way that sanctions the escape of hazardous constituents from the disposal unit suggests that the agency has either misunderstood Congress' directive or is attempting to rewrite the statute to suit its own purposes. But whether the agency's distortion of the statutory language is deliberate or not, it is imperative that the proposed regulations be summarily rejected since they are not consistent with either the letter or the spirit of the law.[28]

Eckart and other critics also complained that EPA was misinterpreting the phrase "for as long as the wastes remain hazardous" as a concept of distance rather than of time.

Eleven House and Senate conferees on the 1984 Amendments formally protested to EPA Administrator Lee M. Thomas that the agency's proposed approach ignored the congressional directive that decisions on treatment of hazardous wastes prior to land disposal be based on a case-by-case petitioning basis, something not included in the EPA proposal. "Including on of a 'no migration' standard," they claimed, "was intended to eliminate land disposal practices that present any threat. It was not intended to allow the continued use of land disposal practices that present threats deemed 'acceptable' by the Agency on the basis of scientific modeling."

According to the conferees, the EPA proposal "totally misses the point" of the 1984 Amendments by ignoring the congressional directive that decisions on land disposal be based on individual petitions. The congressional critics of the agency's proposed approach contended that EPA's modeling assumptions were suspect and unreliable, and that the concept of "maximum acceptable contaminant concentrations" had "no foundation in the law." Procedurally, legally, technically, and in terms of environmental impacts, the congressional hazardous waste leaders strongly protested the EPA proposal, with some saying they would join in lawsuits challenging the approach if it in fact were promulgated without major revisions.

To some, the contretemps over the proposed land disposal rules clearly raised the likelihood of Congress's clamping down even more in the future on the administrative "flexibility" EPA leaders say they need in order to best manage environmental programs. "Congress has been criticized by you and others for the degree of specificity that we included in the HSWA," seven House Democrats and four Senators (two Republicans and two Democrats) wrote Administrator Thomas on March 4, after their subcommittees held hearings on the EPA proposal. "We now find that even with the extraordinary degree of specificity that was included in the HSWA, the agency is determined to ignore clear congressional intent and statutory requirements and to invent its own approach to the problem of hazardous waste land disposal. It is hard to imagine how we could have more clearly expressed our intent."

The early and strident criticisms of its approach were not causing the agency to immediately wave the white flag. Instead, agency officials continued to insist that their proposal was fully protective of human health and the environment and in fact that their modeling assumptions and "screening levels" were so conservative that the vast majority of hazardous wastes would end up being treated prior to disposal in any event. (For their part, industry interests said the agency's modeling assumptions and overall approach were unduly conservative and "overly protective.")

While the agency's top legal advisor, EPA General Counsel Francis Blake, insisted the proposed approach was well within the agency's legal authority under the Amendments, others saw the proposal as, at best, a "stretch" of the statutory language, legislative history, and congressional intent in general. Even those optimistic that the agency's approach could withstand a legal challenge acknowledged the difficulty of demonstrating convincingly that the models and proposed "screening levels" could guarantee safety to human health and the environment. Given the enormous congressional controversy surrounding its proposal, it appeared likely that the agency would have to seriously rethink its policy before proceding to a "final" rule in any event. The "most significant" rule proposed by EPA under the 1984 Amendments in the end would prove also to be the most controversial.

After recalculating its position in light of the heavy barrage of Congressional criticism, EPA, on November 7, 1986 issued a final rule for solvents and dioxin wastes that abandoned the proposed risk-model approach in favor of regulations that set strict treatment standards, based on best demonstrated available treatment.[29] Although EPA imposed an immediate ban on five solvent waste streams[30] unless treated to prevent migration into the environment, the prohibition was delayed until November 8, 1988 for seven others, including dioxin wastes.[31] In addition, the ban was waived for solvent wastes containing less than one percent total solvents or less than one percent total organic carbon, wastes from generators producing less than 100 kilograms per months or wastes from a Superfund clean-up or RCRA corrective action clean-up. In addition, the ban and treatment standards do not apply to solvent wastes in the U and P series (discarded chemical products, intermediates, and off-specification chemicals). EPA's rationale for the delay in the effective dates of the prohibition was based on the lack of commercially available capacity.

Chapter 6 Footnotes

[1] HSWA Sec. 201(a), adding RCRA §3004(g)(6)(B).
References to the various sections of RCRA will be to the sections numbered in the Act. The parallel United States Code citations are as follows:
Section 1004 - 42 U.S.C. §6903
Section 2008 - 42 U.S.C. §6917
Section 3001 - 42 U.S.C. §6921
Section 3002 - 42 U.S.C. §6922
Section 3003 - 42 U.S.C. §6923
Section 3004 - 42 U.S.C. §6924
Section 3005 - 42 U.S.C. §6925
Section 3006 - 42 U.S.C. §6926
Section 3007 - 42 U.S.C. §6927
Section 3008 - 42 U.S.C. §6928
Section 3009 - 42 U.S.C. §6929
Section 3010 - 42 U.S.C. §6930
Section 3012 - 42 U.S.C. §6933
Section 3013 - 42 U.S.C. §6934
Section 3014 - 42 U.S.C. §6935
Section 3015 - 42 U.S.C. §6936
Section 3016 - 42 U.S.C. §6937
Section 3017 - 42 U.S.C. §6938
Section 3018 - 42 U.S.C. §6939
Section 3019 - 42 U.S.C. §6939(a)
Section 7004 - 42 U.S.C. §6974
Section 7006 - 42 U.S.C. §6976
Section 8002 - 42 U.S.C. §6982
Section 9002 - 42 U.S.C. §6991(a)
Section 9003 - 42 U.S.C. §6991(b)
Section 9004 - 42 U.S.C. §6991(c)
Section 9005 - 42 U.S.C. §6991(d)
Section 9006 - 42 U.S.C. §6991(e)
Section 9007 - 42 U.S.C. §6991(f)
Section 9009 - 42 U.S.C. §6991(h)

[2] For a summary of the "deadline litigation," see Quarles, *Federal Regulation of Hazardous Wastes*, Environmental Law Institute, 1982, pp.40-41.

[3] House Committee on Energy and Commerce, Hazardous Waste Control and Enforcement Act of 1983, Rpt. No. 198, 98th Cong., 1st Sess. (henceforth, "House Report") 117 (Minority views).

[4] 130 Cong. Rec. S13816 (daily ed., Oct. 5, 1984).

[5] Adding RCRA Sec. 3004(b)-(n).

[6] HSWA Sec. 101(a), adding RCRA §1002(b)(7).

[7] HSWA Sec. 405(a), adding RCRA §7010(a).

[8] HSWA Sec. 201(a), adding RCRA §3004(c)(1).

[9] Codification, 50 Fed. Reg. 28705 (July 15, 1985).

[10] HSWA Sec. 201(a), adding RCRA §3004(c)(3). This ban is reflected in EPA regulations at 40 C.F.R. §§264.314(f) and 265.314(f). The exception to the liquids in landfills ban allows disposal of nonhazardous liquids at hazardous waste landfills when the only alternative is disposing of the waste at municipal landfills or unlined surface impoundments that contain or might contain hazardous wastes. *See* Codification, 50 Fed. Reg. 28705 (July 15, 1985). In addition, such disposal must not risk contamination of underground drinking water supplies. HSWA Sec. 201(a), adding RCRA §3004(c)(3).

[11] HSWA Sec. 201(a), adding RCRA §3004(e). The dioxin-containing wastes are F020 through F023, and the solvent wastes are those designated as F001 through F005.

[12] HSWA Sec. 201(a), adding RCRA §3004(g)(4)(C). On May 28, 1986, EPA promulgated the schedule for reviewing listed hazardous wastes. 51 Fed. Reg. 19300 (May 28, 1986). In that rule EPA interpreted Section 3004(g)(5)(E) to mean that the hammer and disposal prohibition would not apply to wastes which are *identified* as hazardous but not listed.

[13] Conference Report, Hazardous and Solid Waste Amendments of 1984, Rpt. No. 1133, 98th Cong., 2nd Sess. 86 (henceforth "Conference Report").

[14] HSWA Sec. 201(a), adding RCRA §3004(g)(4).

[15] *Id.*

[16] Conference Report at 88.

[17] HSWA Sec. 201(a), adding RCRA §3004(g)(3).

[18] Conference Report at 87.

[19] HSWA Sec. 201(a), adding RCRA §3004(g)(6)(C).

[20] HSWA Sec. 201(a), adding RCRA §3004(h)(2). Not 30 months as stated in the Conference Report. See also House Report at 87.

[21] HSWA Sec. 201(a), adding RCRA §3004(h)(3).

[22] HSWA Sec. 201(a), adding RCRA §3004(e)(3).

[23] HSWA Sec. 201(a), adding RCRA §3004(j).

[24] HSWA Sec. 201(a), adding RCRA §3004(m).

[25] *Id.*

[26] 51 Fed. Reg. 1602, Jan. 14, 1986.

[27] 51 Fed. Reg. at 1616.

[28] 131 Cong. Rec. H406 (daily ed., Feb. 6, 1986).

[29] 51 Fed. Reg. 40572 (November 7, 1986).

[30] Specially, F001, F002, F003, F004 and F005 that contain more than one percent of 25 separate solvents and that exceed the allowable disposal level for each hazardous constitutent.

[31] Specifically, F020, F021, F022, F023, F026, F027 and F028.

Chapter 7

Minimum Technological Requirements for New Land Disposal Facilities

Summary

Minimum Technological Requirements:

- Double-liner.
- Leachate collection system.
- Groundwater monitoring.

Exceptions:

- Monofills containing foundry wastes.
- Alternative systems that are equally effective.

Required EPA Regulations:

- Siting, by May 8, 1986.
- Design and operating, by November 8, 1986.
- Leak detection, by May 8, 1987.

Groundwater Requirements (New and Existing Facilities):

- Eliminate EPA exceptions to the groundwater monitoring requirement for those facilities above seasonal high water table, having two liners and leachate collection system, and where liners are inspected.
- Retain agency exception to groundwater monitoring requirement of there being no potential for migration of liquids.
- Authorize the agency to grant case-by-case exemptions to groundwater monitoring requirement where the facility does not receive liquid wastes and is designed and operated to exclude precipitation runoff.
- Retroactively apply groundwater monitoring requirement to facilities that received wastes after July 26, 1982.

During the late 1970s and early 1980s, congressional hearings, EPA investigations, and other sources documented numerous instances of groundwater contamination.[1] Although many and varied sources of the pollution were identified, leaking hazardous waste landfills and surface impoundments were quickly determined to be one of the leading causes of groundwater contamination. The search for ways to prevent further contamination of groundwater led Congress in the 1984 RCRA Amendments to specify certain "minimum technological requirements" for land disposal facilities.[2] The level of technical detail specified by Congress in this and in many other provisions of the 1984 RCRA Amendments is unprecedented in environmental legislation.

Specifically, the Amendments establish certain minimum requirements and direct EPA to promulgate even more specific implementing regulations. In addition to developing these regulations, the agency is directed to revisit the issue and initiate subsequent rulemakings to reflect improvements in technological controls and measurements as they occur.[3] EPA estimates that the new technological requirements will lead to additional industry compliance costs of about $40 million to $75 million annually.[4]

Design and Operating Requirements for New Land Disposal Facilities

Under the 1984 Amendments, landfills and surface impoundments (and also expansions and replacements of them) for which permit applications had not been received by November 8, 1984, are deemed to be new, and they must meet certain technology requirements. These requirements, which will be specified in each permit, include:

1) double-liners.

2) leachate collection systems—above and between the liners for landfills and between the liners for surface impoundments.

3) a system of groundwater monitoring.[5]

These requirements also apply to new units or lateral expansions at interim status landfills and surface impoundments for waste received after May 8, 1985.*[6] They do not apply, however, to injection wells, land treatment units, or waste piles.**[7]

Despite the detailed nature of the statutory requirements, EPA in its Codification Rule made the following even more specific interpretations of the new minimum technological provisions:

*Although for the new units, owners and operators had until May 8, 1985, to comply with the minimum technological requirements, they must do so for all those wastes received since the previous November 8. Codification, 50 Fed. Reg. 28707 (July 15, 1985). New units (post November 8, 1984) that ceased receiving wastes by May 8, 1985, are not required to comply with the new minimum technological requirements. *Id.*

**In its Codification Rule, EPA interprets RCRA Section 3015(a) to require interim status waste piles to comply with requirements for a single liner and leachate collection system as specified in existing 40 C.F.R. §264.251(a). 50 Fed. Reg. 28708, n.6 and 28710 (July 15, 1985). The new minimum technological standards of double-liners and a leachate collection system is, however, applicable to any new waste pile unit. Codification, 50 Fed. Reg. 28706, reflected in EPA regulations at 40 C.F.R. §265.254.

- The minimum technological requirements apply not just to new facilities for which permit applications are *received* after the date of the statute's enactment, but also to all facilities for which permits are *issued* after the date of enactment.[8] The effect of that interpretation is to modestly expand the coverage of the provision.
- For purposes of Section 3015's application of the minimum technological requirements at interim status facilities, a new unit, replacement of an existing unit, or lateral expansion of an existing unit is defined as one that first receives waste after November 8, 1984.[9]
- A unit is defined as a contiguous area of land on which waste is placed, or the largest area in which there is a significant likelihood of mixing waste constituents in the same area.[10]
- The liner and leachate collection system must extend to any area of the unit in contact with the waste, including the side walls.[11]

On March 28, 1986, EPA proposed specific design and operating requirements for new landfills, new surface impoundments, replacement of old units and lateral expansions of surface impoundments and landfills. The proposed regulations, which focus on the performance criteria for double liners and leachate collection systems, are intended to prevent the migration of hazardous constituents from the disposal unit during both the active life of the facility and the post-closure care period.[12]

Exception to the Double-Liner and Leachate Collection Requirements

The Amendments allow EPA to waive the double-liner and leachate collection requirements if the owner or operator of the landfill or surface impoundment demonstrates to EPA's satisfaction that an alternative design, together with operating practice and location characteristics, will prevent the migration of hazardous waste constituents into the groundwater or surface water *at least as well* as double-liners and leachate collection systems.*[13] It will be extremely difficult to qualify for this exemption. According to the Senate Report, "there are currently a relatively few facilities located throughout the country which, because of their unique hydrogeological locations and type of operation, may successfully make this demonstration [to qualify for the exemption]."[14] The Senate Report also emphasized that there are numerous uncertainties regarding the transport of hazardous constituents into the environment and that the applicant shoulders the burden of proof in qualifying for the alternative design exemption.[15] Further, as EPA pointed out in its

*EPA is authorized also to grant a waiver for landfills containing only foundry wastes. Adding RCRA Section 3004(o)(3); appearing in EPA regulations at 40 C.F.R. §§264.221(e) and 264.301(e).

Codification Rule, the exemptions provided in the 1984 Amendments would not be available in states, political subdivisions, or localities that did not establish parallel exemptions.[16]

Required EPA Regulations on Design, Siting and Leak Detection

By May 8, 1986, EPA is required to publish guidance criteria identifying areas of vulnerable hydrogeology not suited for the siting of new and existing treatment, storage, or disposal facilities. Then, by November 8, 1987, EPA must establish criteria for the acceptable location of such facilities.[17] Arguing in support of this requirement, Rhode Island Republican Senator John Chafee, chairman of the Senate Subcommittee on Environmental Pollution, stated: "Recent studies show that proper location is at least as important as application of improved technologies. Too many existing facilities are located in areas that aren't well suited to management of hazardous wastes. It is important that these sites be shut down as quickly as possible."[18] According to the Senate Report, the locational criteria should include "such factors as proximity to groundwater or surface waters and, in particular, potential drinking water supplies, . . . wetlands and population concentrations."[19]

By November 8, 1986, EPA must issue regulations or guidance documents implementing the minimum design and operating requirements for new landfills and surface impoundments.[20] In the meantime, an owner or operator can satisfy the double-liner requirement by installing a top liner capable of preventing migration of any constituents into the liner during the facility's life, including any post-closure monitoring period, and an impermeable lower liner of clay or other natural material.*[21]

By May 8, 1987, EPA must promulgate standards for leak detection systems for new landfill units, surface impoundment units, waste piles, underground tanks and land treatment units. The term "new units" means those on which construction begins after the date of publication of the leak detection regulations.[22] The provision requiring the use of leak detection was added by an amendment offered by Congressman Billy Tauzin (D-Louisiana). Tauzin stressed that his amendment would help avoid pollution before it occurs, rather than merely cleaning it up after the fact. Under EPA's groundwater requirement, he pointed out, operators "are supposed to check the water to see if it is already polluted. . . . My amendment . . . would allow us to discover toxic leaks before they get into the water supply."[23]

*The agency apparently believes that the statutory requirement of preventing "the migration of any constituent" during the life of the facility was inadequate for the interim period prior to the issuance of the minimum design and operating requirements. Consequently, EPA has decided to use its discretion to require that the liner system be sufficient also to protect human health and the environment. Specifically, the agency said in its Codification Rule that it "believes that, during the interim period, an owner or operator who wishes to install a liner system other than the one described in Section 3004(o)(5)(B) [lower liner being no more permeable than 1×10^7 centimeters per second] must meet the broad narrative standard of protection of human health and the environment." 50 Fed. Reg. 28709 (July 15, 1985).

It should be noted that the leak detection rule is intimately related to EPA action establishing minimum technological requirements for leachate collection. In its Codification Rule, EPA stated that "the legislative history suggests that the leachate collection system should act both as a leachate collection and removal system and a leak detection system."[24]

Groundwater Monitoring Requirements (New and Existing Facilities)

Section 203 eliminates certain EPA exceptions to the RCRA groundwater monitoring requirements that Congress believed were unwarranted.[25] Groundwater monitoring now is required for the following situations which EPA previously had exempted by regulation: (1) where the facility is located above the seasonal high water table; (2) where the facility has two liners and a leachate collection system; and (3) where the liners are inspected.

This provision specifically states that it does not affect other regulatory exemptions for groundwater monitoring that were in effect on the date of enactment of the 1984 Amendments. EPA interprets this to mean that only two groundwater monitoring waivers remain available.[26] The first waiver allows a groundwater monitoring exemption for units posing no potential for migration of liquids.[27] The second waives groundwater monitoring for land treatment units if it is demonstrated that no hazardous constituents have migrated beyond the treatment zone during the active life of the unit.[28] Section 203 also authorizes EPA to grant additional exemptions on a case-by-case basis for structures which do not receive liquid wastes and which are designed and operated in a way that excludes precipitation runoff, so long as they meet other stringent technological criteria.[29]

One of the most important changes that the Amendments make with respect to groundwater monitoring is the "retroactive" application of groundwater monitoring requirements and corrective action requirements to facilities that received wastes after July 26, 1982.[30] Previously, only facilities that had received waste after January 26, 1983, had to comply with these requirements. EPA had issued the requirements on July 26, 1982, to be effective on January 26, 1983. This gave problem facilities a six-month period to close and thereby avoid groundwater monitoring and corrective action. Many facilities were taking advantage of the six-month loophole, so Congress decided to close it. According to the House Report, this provision was adopted "for the purpose of correcting one of the most serious deficiencies" in EPA's regulatory program.[31]

Minimum Requirements for Incinerators

Section 202(a) requires hazardous waste incinerators to meet the 99.99 percent minimum destruction and removal efficiency (DRE) standards established by EPA on June 24, 1982.[32] By making this a statutory requirement, Congress wanted to prevent EPA from backing off its own established position, which it was considering doing during the Lavelle regime by the adoption of a cost/risk analysis scheme.[33] The House Report expressly disagreed with EPA's intention of developing standards based on cost/risk analysis, which some members of the House Energy and Commerce Committee viewed as the agency's rationale not only for weakening the DRE standard but also for undermining RCRA's fundamental mandate to protect human health and the environment.[34] It should be noted that this DRE statutory requirement currently applies only to incinerators, not to other types of combustion units such as boilers and industrial furnaces, which are to be regulated under Section 3004(q).[35]

Chapter 7 Footnotes

[1] See Congressional Research Service *Groundwater Contamination by Toxic Substances: A Digest of Reports*, S. Rpt. 98-131 (November 1983); Council on Environmental Quality, *Contamination of Ground Water by Toxic Organic Chemicals*, (January 1981); EPA, *The Prevalence of Subsurface Migration of Hazardous Chemical Substances at Selected Industrial Waste Land Disposal Sites* (1977).

[2] House Report at 62; Senate Committee on Environment and Public Works, Solid Waste Disposal Act Amendments of 1983, Rpt. No. 284, 98th Cong., 1st. Sess. 26 (henceforth "Senate Report").

[3] HSWA Sec. 202(a), adding RCRA §3004(o)(1).

[4] Congressional Budget Office, *Hazardous Waste Management: Recent Changes and Policy Alternatives*, 53, Table 16 (May 1985).

[5] HSWA Sec. 202(a), adding RCRA §3004(o)(1)(A). The double-liner and leachate collection system are reflected in EPA's regulations at 40 C.F.R. §264.221(c) (surface impoundments) and 40 C.F.R. §301(c) (landfills).

[6] HSWA Sec. 243(a), adding RCRA §3015(b).

[7] Senate Report at 27 and HSWA Sec. 243(a), adding RCRA §3015(a).

[8] 50 Fed. Reg. 28708 (July 15, 1985).

[9] 50 Fed. Reg. 28707.

[10] 50 Fed. Reg. 28706.

[11] 50 Fed. Reg. 28709.

[12] 51 Fed. Reg. 10719, 10720 (Mar. 28, 1986).

[13] HSWA Sec. 202(a), adding RCRA §3004(o)(2), appearing in EPA regulations at 40 C.F.R. §§264.221(d) and 264.301(d).

[14] Senate Report at 27 and 28.

[15] Senate Report at 27.

[16] 40 Fed. Reg. 28729-30 (July 15, 1985).

[17] HSWA Sec. 202(a), adding RCRA §3004(o)(7).

[18] 130 Cong. Rec. S13819 (daily ed., Oct. 5, 1984).

[19] Senate Report at 30.

[20] HSWA Sec. 202(a), adding RCRA §3004(o)(5)(A).

[21] HSWA Sec. 202(a), adding RCRA §3004(o)(5)(B).

[22] HSWA Sec. 202(a), adding RCRA §3004(o)(4)(B)(ii).

[23] 130 Cong. Rec. H8150 (daily ed., Oct. 6, 1983).

[24] 50 Fed. Reg. 28709 (July 15, 1985).

[25] Senate Report at 64.

[26] Codification, 50 Fed. Reg. 28717 (July 15, 1985).

[27] 40 C.F.R. §264.90(b)(4).

[28] 40 C.F.R. §264.280(e).

[29] This exemption has been incorporated in EPA regulations at 40 C.F.R. §264.90(b)(2).

[30] HSWA Sec. 243(c), adding RCRA §3005(i); reflected in EPA regulations at 40 C.F.R. §264.90(a).

[31] House Report at 44.
[32] Adding RCRA §3004(o)(1)(B).
[33] House Report at 62.
[34] House Report at 62 and 63.
[35] House Report at 42.

Chapter 8

Interim Status Land Disposal Facilities: Minimum Technological Requirements and Changes in Status

Summary

Minimum Technological Requirements:

• Existing surface impoundments must install double-liners and leachate collection systems by November 8, 1988.
• Existing landfills not required to retrofit.

Exceptions for Surface Impoundments:

• Facility located more than one-fourth mile from drinking water aquifer and has a liner that is not leaking.
• Facility treating wastewater during the secondary or tertiary phase of aggressive biological treatment.
• Facility where there will be no migration of any hazardous constituent.

Application for Exception Must Filed by November 8, 1986, and Must Contain:

• Groundwater monitoring data.
• Evidence of whether impoundment is leaking.
• Certification from a registered professional engineer.

Loss of Interim Status for Land Disposal Facility, Unless by November 8, 1985, Applicant:

• Applies for a permit.
• Certifies compliance with groundwater monitoring and financial responsibility requirements.

Expansion of Interim Status Category to Include Facilities Covered by Subsequent Regulations, where Applicant:

• Applies for a permit within one year of coverage.
• Complies with notification requirements of Section 3010.

Design and Operating Requirements for Surface Impoundments

One of the most dramatic changes mandated by the 1984 Amendments is the retrofit requirement for surface impoundments containing hazardous wastes. Congressman George Brown, Democrat of California, pointed out during the reauthorization debate a statistic that surprised many legislators—that 70 percent of surface impoundments are unlined.[1] EPA now believes that 90 percent of the nation's hazardous waste surface impoundments are designed and/or located in a manner that could result in groundwater contamination.[2]

Section 215 of the Amendments addresses this problem by requiring that existing units at interim status surface impoundments comply with the new minimum technological requirements for new facilities (double-liners, leachate collection system, groundwater monitoring) by November 8, 1988. If they cannot comply, they must stop receiving, storing, or treating hazardous wastes.[3] Existing landfills are not subject to the new technological requirements, in part because EPA and others contended that the process of retrofitting these facilities actually might cause more environmental harm than it would correct.[4]

Existing surface impoundments that become subject to the technological requirements by subsequent regulations—including those covered as the result of the listing of additional wastes as hazardous—must comply within four years of the effective date of those subsequent regulations. Impoundments that are initially excluded by an exception but later determined to be covered must comply within two years of the effective date of the new determination, except for wastewater impoundments, which have three years to comply.[5]

Where it determines that hazardous wastes are likely to migrate to groundwater, EPA may apply the new technological requirements, or other protective measures, *at any time* in order to protect human health and the environment.[6] This provision amounts, in practice, to an override of the exemptions to the technological requirements discussed below. The new technological requirements must be met immediately at surface impoundments where EPA allows an extension of the land restriction deadlines.[7]

Exceptions to the Liner and Leachate Collection Requirements

Section 215 provides three principal exceptions to the new technological liner and leachate collection requirements for interim status surface impoundments.* It should be emphasized, however, that the exceptions are not available if the state in which the facility is located has not established parallel exemptions.

The first exception is for any surface impoundment which (1) has at least one liner which is not leaking; (2) is located more than one-fourth of a mile from an underground source of drinking water (Only about 5 percent of the impoundments fall into this category, according to EPA.); and (3) is in compliance with groundwater monitoring requirements for permitted facilities.[8]

*Additionally, EPA interprets the two exceptions for new facilities as applicable to interim status surface impoundments. Codification, 50 Fed. Reg. 28711 (July 15, 1985).

At the time of closure of any surface impoundment exempted on this basis, the owner or operator must remove or decontaminate all waste residues, all contaminated liner material, and all contaminated soil to the extent practicable. If removal of contaminated soil is not practicable, the owner or operator must comply with all applicable post-closure requirements.[9]

A second exception applies to surface impoundments treating wastewater during the secondary or tertiary phase of "aggressive" biological treatment. To be eligible, the surface impoundment must be in compliance with all applicable groundwater monitoring requirements. It also must be a part of a facility with a Clean Water Act permit and must meet all applicable Clean Water Act regulations.[10]

According to Senator John H. Chafee (R-Rhode Island), this exemption was needed in order to avoid interfering with water pollution treatment methods that rely on surface impoundments, and to preclude potential conflicts with effluent limitations promulgated by EPA under the Clean Water Act.[11] Senator Lloyd Bentsen (D-Texas) amplified on certain key terms of the exception, noting that secondary and tertiary treatment refer to biological treatment rather than the second and third ponds of a multi-impoundment system. Bentsen said the word "aggressive" was used in order "to distinguish between [facilities] where biological activity is an incidental rather than a primary purpose of the impoundment system."[12]

Section 215 also provides that, if a surface impoundment eligible for this second exemption subsequently is found to be leaking, EPA must require compliance with the new technological requirements unless the agency determines that such compliance is not necessary to protect human health and the environment.[13] Congress was well aware that this exemption would be the subject of continuing controversy. Consequently, it directed EPA to submit to it a report on the environmental consequences of the exemption and on the feasibility and cost of deleting it. Unlike virtually all of the other studies required by the 1984 Amendments, no deadline for the report on the wastewater exemption was established.[14]

A third exception allows EPA to modify the technological requirements where the applicant demonstrates that there will be no migration of any hazardous constituent from the impoundment at any future time.[15] Satisfying this test is virtually impossible, but it responds to arguments made to Congress that no tighter technological measures were necessary because migration would not occur in many situations. Senator Jeremiah Denton (R-Alabama) attempted to ease the burdens imposed by this exception with an explanation that "any future time" was not intended to be a legal bar to EPA's granting the exception, and that 150 years falls within the meaning of the phrase.[16] However, Congressman John Breaux (D-Louisiana), rejected Denton's 150-year interpretation, stating that it did not represent the intent of the House-Senate conferees, nor his own intent as the original author of the provision.[17] "The phrase 'any future time' means exactly what it says," Breaux insisted.[18]

There are two other situations in which the minimum technological requirements can be modified for surface impoundments. The first involves liners and leak detection systems installed in "good faith compliance" with the then-current technological requirements for new facilities. In this situation, EPA could not require a different liner or leak detection system in the first permit issued for the facility. However, EPA may require a new liner later if it has reason to believe the existing liner is leaking.[19] EPA interprets this authority very narrowly, stating that it can require retrofitting only at the time of permitting.[20]

A second situation in which the minimum technological requirements may be waived involves a surface impoundment which is undertaking corrective action in compliance with a consent decree or agreement entered into before October 1, 1984. To be eligible for the exemption, however, the impoundment's corrective action system must provide a degree of protection at least equivalent to the minimum technological requirements.[21] This amendment was offered by Representative Barbara Mikulski (D-Maryland), to assist a company in her district which owned a surface impoundment that was under court order to proceed with a pump and treat corrective action procedure. However, according to EPA, it is not clear that the exemption as written will actually help Mikulski's constituent.

Application for Exception

An application for any of the three exceptions must be filed by November 8, 1986, and must include the following: a permit application, groundwater monitoring data, all reasonably ascertainable evidence on whether the impoundment is leaking, and a certification by a registered professional engineer "with academic training and experience in groundwater hydrology" that the impoundment meets the applicable criteria.[22] In addition, EPA must provide notice of and opportunity to comment on an application for an exception and then rule on it within 12 months of its receipt.[23]

Expansion of Interim Status Category

Section 3005(e) of RCRA previously restricted interim status to owners or operators of hazardous waste management facilities in existence on November 19, 1980. The 1984 Amendments modestly expand eligibility for interim status by making it available to facilities that are in existence on the effective date of a statutory or regulatory change which makes that type of facility subject to a RCRA permit. To qualify for interim status under this provision, the facility must apply for a permit and comply with the Section 3010 notification requirements.[24] Interim status is not available, however, for facilities previously denied a RCRA permit, nor for facilities for which authority to operate under RCRA has been previously terminated.[25]

Loss of Interim Status Designation

It is somewhat ironic that Congress in enacting the 1980 Amendments ratified and expanded EPA's interim status program, but four years later made certain that the program would be phased out as quickly as possible. This change of attitude reflects an overall frustration with the extraordinary delays in the permitting process, and also a belief that the owners of many facilities had no intention of coming into compliance, but rather would close their facilities only if EPA ever got around to requiring submission of their Part B (i.e., final permit) applications. Thus, the Amendments mandate that interim status terminate for land disposal facilities unless, by November 8, 1985, the owner or operator submits a Part B permit application and certifies compliance with groundwater monitoring and financial responsibility requirements (discussed in detail in Chapter 18).[26] For facilities that became subject to the permit requirement by a subsequent statutory or regulatory requirement, interim status expires one year after the new requirement takes effect, unless the owner or operator makes a permit application and submits the certification.[27] Previously, under EPA's regulations, interim status could continue indefinitely; applications for final permits were submitted either voluntarily or in response to a request from an authorized state or EPA.[28]

For incinerators for which a permit application was submitted before the 1984 Amendments were enacted, interim status terminates on November 8, 1989, unless the owner or operator applies for a final determination regarding the issuance of a permit by November 8, 1986.[29] On November 8, 1992, interim status terminates for facilities other than land disposal units or incinerators for which a permit was applied before the Amendments, unless the owner or operator applies for a final determination regarding the issuance of the permit by November 8, 1988.[30]

Chapter 8 Footnotes

[1] 129 Cong. Rec. H8149 (daily ed., Oct. 6, 1983).

[2] Congressional Budget Office, *Hazardous Waste Management: Recent Changes and Policy Alternatives*, 2 (May 1985).

[3] Adding RCRA §3005(j)(1). On March 28, 1986, EPA proposed rules concerning double-liners. 51 Fed. Reg. 10706 (Mar. 28, 1986). In its discussion of double-liners, the Agency stated that it did "not believe that the interim bottom liner design of Section 3004(o)(5)(B), i.e., three feet of compacted soil material with a hydraulic conductivity of 1×10^{-7} centimeters per second will, in most cases, meet the bottom liner performance standard of preventing the migration of any hazardous constituent through it during the active life and post-closure care period. 51 Fed. Reg. at 10709.

[4] House Report at 45.

[5] HSWA Sec. 215, adding RCRA §3005(j)(6).

[6] HSWA Sec. 215, adding RCRA §3005(j)(7)(B).

[7] HSWA Sec. 215, adding RCRA §3005(j)(11).

[8] HSWA Sec. 215, adding RCRA §3005(j)(2).

[9] HSWA Sec. 215, adding RCRA §3005(j)(9).

[10] HSWA Sec. 215, adding RCRA §3005(j)(3).

[11] 130 Cong. Rec. S13819 (daily ed., Oct. 5, 1984).

[12] 130 Cong. Rec. S9183 (daily ed., July 25, 1984).

[13] HSWA Sec. 215, adding RCRA §3005(j)(7)(C).

[14] HSWA Sec. 215, adding RCRA §3005(j)(7)(A).

[15] HSWA Sec. 215, adding RCRA §3005(j)(4).

[16] 130 Cong. Rec. S13812 (daily ed., Oct. 5, 1984).

[17] 130 Cong. Rec. H4455 (daily ed., Oct. 10, 1984).

[18] *Id.*

[19] HSWA Sec. 215, adding RCRA §3005(j)(8); reflected in EPA regulations at 40 C.F.R. §§265.221(e) and 265.301(e).

[20] Codification, 50 Fed. Reg. 28708 (July 25, 1985).

[21] HSWA Sec. 215, adding RCRA §3005(j)(13).

[22] HSWA Sec. 215, adding RCRA §3005(j)(5).

[23] *Id.*

[24] HSWA Sec. 213, amending RCRA §3005(e).

[25] These new statutory provisions on interim status are now reflected in EPA's regulations at 40 C.F.R. §270.70(a) and (c).

[26] HSWA Sec. 213, adding RCRA §3005(e)(2)(A).

[27] *Id.*, adding RCRA §3005(e)(3).

[28] 40 C.F.R. §270.10(e)(4).

[29] HSWA Sec. 213, adding RCRA §3005(c)(2)(C).

[30] *Id.* The various statutory termination dates for interim status are reflected in the EPA regulations at 40 C.F.R. §270.73.

Chapter 9

Permits

Summary

Permit Life:

• Fixed term not to exceed 10 years for land disposal, storage, incineration, and other treatment.
• Land disposal permits must be reviewed and modified as necessary every five years.

Permit Issuance Timetable:

• Land disposal facilities by November 8, 1988.
• Incinerators by November 8, 1989.
• Other facilities by November 8, 1992.

Permit Issuance Requires Contamination Be Cleaned Up:

• Corrective action to clean up contamination required for all units of solid waste management facility before any unit can receive a permit.
• Corrective action applies both to post-closure permits and to operating permits.
• Current owners responsible for releases from wastes received by past owners.

Permit Requires Exposure Assessment which Addresses:

• Potential for releases from normal operations and transportation accidents.
• Potential pathways of human exposure resulting from such releases.
• Potential magnitude and nature of the human exposure.

Preconstruction Ban/TSCA Exception:

• Permit required for construction of new hazardous waste management facility.
• Exception for PCB facility which requires just operating permit.

Because most of Congress' hazardous waste policy objectives can be carried out only through the issuance and enforcement of permits, it is not surprising that Congress made some major revisions to this aspect of RCRA. Among the changes are the establishment of deadlines by which EPA must issue permits, a requirement that a permit application be made by November 8, 1985, for retention of interim status, the creation of permits to encourage innovative treatment technology, and a ban on construction of new hazardous waste management facilities pending issuance of a permit.

Permit Life

On May 19, 1980, EPA promulgated regulations providing that RCRA permits would be effective for no more than 10 years.[1] However, the agency on February 8, 1983, proposed to amend that rule so it could issue permits for the life of a facility.[2] The proposal was not made final because, before EPA could act, Congress did. Section 212 of the 1984 Amendments provides that any permit for land disposal, storage, incineration, and other treatment shall be for a fixed term of no more than 10 years.[3] By way of comparison, discharge permits issued under the Clean Water Act have a five-year duration.

The rationale for a 10-year permit life is contained in the Senate Report, which points out that "with the advancing state of technology and the long projected useful life of many of these facilities, it is preferable to limit permit life to the minimum period consistent with the cost and administrative burden of issuing a permit. . . . Limited permit duration will assure that facilities are periodically reviewed and requirements for them upgraded to reflect the current state of the art."[4] The agency regulation, as it existed prior to the Amendments,[5] is consistent with the statutory provision on permit life and therefore was not amended by EPA's Codification Rule.

Section 212 also requires that EPA review land disposal permits every five years and modify permits to assure that the facilities continue to meet the currently applicable requirements.[6] Under this approach, EPA must specifically focus on whether individual permits meet evolving standards. The Senate Report states: "In conducting such reviews and in deciding whether or not to modify the permit, the agency or the state shall consider any changes that may have occurred in the operation of the facility since the permit was issued, standards and requirements of current regulations under Sections 3004 and 3005, advances in hazardous waste control practices and technology since permit issuance, and other information concerning the impact of the facility on human health and the environment."[7] Also, the 1984 Amendments authorize, but do not require, EPA to review and modify any permit at any time during its term to reflect improvements in the state of control and measurement technology, or to reflect changes in applicable regulations.[8]

Permit Issuance Timetable

The imposition of deadlines for issuing permits responds to the snail's pace at which EPA had proceeded with permitting activities. As of July 31,

1983, final permits had been issued to only 24 facilities out of approximately 8,000. Of these 24 final permits, one had been issued to a landfill, three to incinerators, and the rest to storage facilities.[9]

With the enactment of the 1984 Amendments, permits for land disposal facilities are required to be issued by November 8, 1988, for incinerators by November 8, 1989, and for other facilities by November 1992.[10] The House bill would have required that EPA give highest priority to processing permits for facilities currently contaminating groundwater. The Conference Report modified this requirement, however, stating that priority should be given to issuing permits to qualified facilities and to those contaminating groundwater.[11] Thus, the Amendments place as much importance on identifying facilities to which hazardous wastes can be safely sent as on identifying facilities for which corrective action is necessary.

In August of 1985, EPA stated that it expected to meet the deadlines for issuance of incinerator and storage permits, but that it "is going to be very difficult to meet" the deadline for land disposal facilities.[12]

Requirement of Corrective Action for Permit Issuance

Section 206* mandates that permits issued after enactment of the 1984 Amendments for treatment, storage, and disposal facilities must require corrective actions for all releases of hazardous constituents from any *solid waste* management unit at a facility, regardless of the time the waste was placed there. Where corrective action cannot be completed prior to issuance of the permit, the permit must contain a compliance schedule and assurances of financial responsibility for completing the corrective action.[13] In its Codification Rule, EPA interprets this provision to apply both to post-closure permits and operating permits.[14] Further, it interprets "release" to encompass release to all media including groundwater[15] and holds current owners of the facilities responsible for releases from wastes received by previous owners.[16]

The corrective action provision has been referred to as a "sleeper" and a "mini-Superfund," because it will require large financial expenditures to accomplish what may involve a massive clean-up effort.** Yet the legislative history left little doubt that Congress intended broad application of the provision. The House Report, for instance, explicitly stated: "The purpose of this provision is to ensure that all facilities which seek a permit under 3005(e) control and clean up all releases of all hazardous constituents from all solid waste management units at the time of permitting the facility. . . . The Committee believes that all facilities receiving permits should be required to clean up all releases from all units at the facility, whether or not such units are cur-

*This provision was known as the "Prior Releases" provision in the House Report and the "Continuing Releases" provision in the Senate and Conference Reports.

**An EPA survey completed in April 1986 revealed that 50 to 60 percent of all storage facilities have solid waste management units (SMUs) and 80 to 90 percent of all land disposal facilities have SMUs. On average, according to the survey, a RCRA-regulated facility has six SMUs. *Hazardous Waste News,* April 28, 1986, at 165.

rently active."[17] Congress believed it would make little sense for a facility to comply with all the new technology requirements to prevent groundwater contamination at a new unit and, at the same time allow the old units to continue to pollute.

The agency has softened the potential impact of this provision by stating in its Codification Rule that it will only mandate corrective action under the provision "where necessary to protect human health and the environment."[18] Nonetheless, many generators now retaining their wastes at the sites probably will ship them to commercial, off-site facilities to avoid the impact of the corrective action provision.

In addition to requiring corrective action as part of permitting, the Amendments also authorize EPA to require corrective action at an interim status facility whenever it determines there is or has been a release of hazardous waste into the environment.[19] In this case, corrective action is discretionary with EPA.

Exposure Assessment Requirement for Permits

Congressman Estaban Torres, a California Democrat whose thirty-fourth district includes the infamous BKK landfill, offered an amendment on the House floor on October 6, 1983, that was aimed at focusing EPA's attention on the leaks from the BKK facility and the need to give people living near hazardous waste landfills the right to know if the facilities posed a threat to their health. According to Congressman Torres, "[W]ithout the facts, our constituents have only to fear the worst. . . ."[20] Congressman Don Ritter (R-Pennsylvania) opposed the exposure assessment amendment. While acknowledging that the proposal was "well intentioned," he argued that the studies would be very expensive to conduct and would result in data "that will scare people, that may or may not be true, that the politicians will run around with and flail before the television cameras, and that the media will just jump up and down with."[21] Despite Ritter's grave concerns, the House adopted the Torres amendment.

The provision, which was substantially modified in the Conference Committee, required that by August 8, 1985, permit applications for landfills and surface impoundments (but not other disposal, storage, or treatment facilities) must be accompanied by information on the potential for public exposure to hazardous substances released from those units. The information must be "reasonably ascertainable by the owner or operator,"*[22] and must address at a minimum:

- reasonably foreseeable potential releases from both normal operations and accidents including releases associated with transportation;
- the potential pathways of human exposure resulting from such releases;

*Congressman Torres stated that he "would anticipate that a landfill operator will hire an independent consultant to conduct the health study." 129 Cong. Rec. H8155 (daily ed., Oct. 6, 1983).

- the potential magnitude and nature of the human exposure.[23]

Facilities whose applications had been submitted to EPA were given until August 8, 1985, to submit this exposure information.

If the exposure information indicates the presence of a substantial risk, EPA may request that the Agency for Toxic Substances and Disease Registry conduct a health assessment.[24] Funds for the health assessments can be recovered from the owner of the facility under Section 107 of CERCLA for facilities found to be contributing to a release of a hazardous substance.[25]

In its Codification Rule, EPA explained that the exposure information is not actually a part of a permit application, but rather that it must accompany the application.[26] The agency further stated that Congress in making this distinction intended to prevent the preparation and submission of exposure information from delaying the permit process.[27] Thus, noncompliance with the health assessment requirement would be a separate violation of RCRA, not a matter to be considered when determining whether a complete permit application had been submitted.

Preconstruction Ban/PCBs Exception

Prior to the 1984 Amendments, EPA had adopted regulations prohibiting the construction of new hazardous waste management facilities not having a final RCRA permit.[28] However, the prohibition had been left in doubt as the result of litigation, and Congress felt it was necessary to clarify EPA's authority to require a RCRA permit as a prerequisite to constructing a hazardous waste treatment, storage, or disposal facility. The agency's rationale for the prohibition was simple: it wanted to assure that all new facilities would be properly designed, constructed, and located.

The preconstruction ban, however, is qualified as to polychlorinated biphenyls (PCBs). PCBs are not currently listed as a hazardous waste, and they therefore are not subject to RCRA requirements, but rather to the Toxic Substances Control Act (TSCA). The Conference Report, however, urges EPA to list PCBs as expeditiously as possible.[29] The Amendments provide that, if PCBs are listed as a hazardous waste, no RCRA preconstruction permit will be required for their incineration if the facility has a permit for incineration of PCBs under Section 6(e) of TSCA.[30] That provision of TSCA does not require preconstruction approval. The facility will, however, be required to obtain a RCRA operating permit subsequent to construction.[31]

The Amendments did not specifically address whether the TSCA exemption from the preconstruction ban applies to units that store PCBs prior to incineration at a TSCA-approved incineration facility. In its Codification Rule, EPA interprets the Amendments to provide such an exemption.[32]

New and Innovative Technology

On October 6, 1983, an amendment designed to encourage development of new and innovative treatment technology was offered on the House floor by Congressman Ike N. Skelton, a Missouri Democrat whose Small Business Subcommittee had held hearings on the topic. As a result of his subcommittee's investigations, Skelton had concluded that, although the nation certainly needed innovative approaches to solve the growing problems of hazardous waste disposal, the current RCRA statute and EPA regulations impeded such solutions.[33] "Innovators must go through a permitting process designed for operational hazardous waste facilities before they can even experiment with a new technology or demonstrate its feasibility," Skelton said. "Furthermore, in many cases EPA has not yet promulgated technical standards to cover new and innovative technologies so that the permit cannot be issued."[34]

Under Skelton's new technology provision, included in the 1984 Amendments after slight revision by the Conference Committee,[35] EPA is authorized to issue research, development and demonstration permits for facilities which propose to use experimental treatment technology or processes not presently governed by EPA regulations. These permits are limited to one year, but they can be renewed three times.

The new technology provision makes clear that EPA is not required to promulgate general regulations on experimental permit criteria, but that it must include in individual permits requirements that adequately protect human health and the environment. EPA is authorized to waive general requirements for permits, except those dealing with financial responsibility and public participation. Also, EPA is explicitly authorized to terminate the operation of an experimental facility at any time in order to protect human health.[36]

The legislative history provides three examples of the types of research, development, and demonstration activities Congress intended to be covered by this provision.[37] According to EPA, those examples suggest some congressional limitations on agency authority to issue these permits.[38]

In the first example, an individual or company has designed on paper or in the laboratory an innovative system for treating hazardous waste. But, in order to determine whether this new technology is technically feasible, a small pilot scale unit must be constructed and operated. The second example involves an equipment vendor and a waste-generating or -processing customer. Vendors often custom-prepare storage and processing equipment, such as tanks and incinerators, based on a customer's individual needs. This may require one or more tests with a pilot facility using samples of the customer's waste. In the third example, a manufacturer or user of a particular commercial treatment process wants to improve its efficiency or effectiveness or to reduce environmental hazards. Doing so may involve construction of a pilot scale treatment unit that will be operated on an experimental basis to test new wastes or different operating conditions.

Despite its statement in the Codification Rule that the examples imply limitations on research, development and demonstration permits, EPA hedges on the issue by stating that the examples are not an exclusive list of activities permissible under the provision.[39] Also in the Codification Rule, EPA notes that if the waste management activity operating under such a permit is used at any time to store or treat waste for any reason other than the conduct of a treatment experiment, it must be permitted and operated in accordance with all applicable requirements of 40 C.F.R. Parts 264 and 266.[40]

An important limitation for research, development and demonstration projects is that, in states authorized to administer the permit program, the state permitting requirements will apply to such projects unless the state has an analogous "fast track" provision.[41] Thus, in those states without the analogous provisions, even if a facility obtains a research, development and demonstration permit from EPA, it still would not be allowed to operate until it has obtained a state permit imposing general requirements. Obviously, this result would negate the whole purpose of the special provision. The solution to this problem is for the states to either modify their programs to allow for operation under the federal permit alone, or adopt permit provisions analogous to the federal standard.

In July 1985, EPA proposed to issue the first research permit under the new statutory provision to the Atlantic Research Corporation of Virginia for a project to study the decontamination of soil by composting. Under the proposed permit, the company will collect soil containing potentially hazardous wastes from two military disposal sites containing trinitrotoluene (TNT) and other toxic substances. The proposed study will examine the effect of microbial action on pollutants by soil bacteria.[42]

Chapter 9 Footnotes

[1] 40 C.F.R. §122.9(a), now codified as 40 C.F.R. §270.50(a).
[2] 47 Fed. Reg. 5872.
[3] HSWA Sec. 212, adding RCRA §3005(c)(3).
[4] Senate Report at 30.
[5] 40 C.F.R. §270.50(a)-(c).
[6] HSWA Sec. 212, adding RCRA §3005(c)(3). The agency has incorporated this requirement into its regulations at 40 C.F.R. §270.50(d).
[7] Senate Report at 31.
[8] HSWA Sec. 212, adding RCRA §3005(c)(3).
[9] GAO, Interim Report on Inspection, Enforcement, and Permitting Activities at Hazardous Waste Facilities, 18 (Sept. 21, 1983).
[10] HSWA Sec. 213, adding RCRA §3005(c)(2).
[11] Conference Report at 95.
[12] Letter of Aug. 19, 1985, from J. Winston Porter, EPA Assistant Administrator for Solid Wastes and Emergency Response, to Senator Robert T. Stafford.
[13] HSWA Sec. 206, adding RCRA §3004(u).
[14] Codification, 50 Fed. Reg. 28712 (July 15, 1985).
[15] Codification, 50 Fed. Reg. 28713.
[16] Codification, 50 Fed. Reg. 28714. The new requirements imposed by RCRA §3004(u) and EPA's interpretations are contained at 40 C.F.R. §264.101.
[17] House Report at 60; see also Senate Report at 3.
[18] Codification, 50 Fed. Reg. 28713. On March 28, 1986, EPA proposed rules governing corrective action for releases of hazardous waste or hazardous constituents from solid waste management units. The proposal would amend the Part B permit application to require owner/operators to submit two types of information: a full description of the solid waste management unit and all available data pertaining to any release from the unit. 51 Fed. Reg. 10713.
[19] HSWA Sec. 233, adding RCRA §3008(h).
[20] 129 Cong. Rec. H8155 (daily ed., Oct. 6, 1983).
[21] 129 Cong. Rec. H8157 (daily ed., Oct. 6, 1983).
[22] HSWA Sec. 247, adding RCRA §3019(a).
[23] *Id.*
[24] HSWA Sec. 247, adding RCRA §3019(b)(2).
[25] HSWA Sec. 247, adding RCRA §3019(g). The statutory requirements for exposure assessments are reflected in EPA regulations at 40 C.F.R. §270.10(j).
[26] Codification, 50 Fed. Reg. 28726.
[27] *Id.*, citing 130 Cong. Rec. S9187 (daily ed., July 25, 1984).
[28] 40 C.F.R. §270.10(f)(1).
[29] Conference Report at 105.
[30] HSWA Sec. 211, amending RCRA §3005(a).

[31] EPA has incorporated this PCBs exemption in its regulations at 40 C.F.R. §270.10(f)(3).

[32] Codification, 50 Fed. Reg 28722 (July 15, 1985).

[33] 129 Cong. Rec. H8159 (daily ed., Oct. 6, 1983).

[34] *Id.*

[35] HSWA Sec. 214(a), adding RCRA §3005(g).

[36] *Id.*

[37] 129 Cong. Rec. H88160 (daily ed., Oct. 6, 1983).

[38] Codification, 50 Fed.Reg. 28728 (July 15, 1985).

[39] *Id.*

[40] *Id.*

[41] Codification, 50 Fed.Reg. 28729 (July 15, 1985).

[42] Hazardous Waste Report, 10, (July 22, 1985).

Chapter 10

Small Quantity Generators

Summary

EPA Regulations:

- Must regulate generators of wastes in quantities between 100 and 1,000 kilograms per month.
- May regulate generators of less than 100 kilograms of waste per month.
- May vary the standards from those for large quantity generators.
- Must at a minimum require treatment and disposal of hazardous wastes from small quantity generators at permitted or interim status facilities.
- May allow storage on the site without a permit for up to 180 days and in some cases 270 days.

Hammer Provisions:

- Small quantity generator regulations must be issued by March 31, 1986, and if not,
- Wastes can be disposed of only at permitted or interim status facilities.
- The interim manifest requirement applies and the name of the waste transporter must be included.

Reports and Publicity:

- Various reports on small quantity generators must be issued by April 1, 1985.
- For 30 months after enactment of the Amendments, EPA must inform small quantity generators of their new statutory responsibilities.

The Senate Report specified that the small quantity generator amendment is intended to correct a current regulatory exclusion from the Subtitle C program that was not contemplated by Congress in enacting the 1976 Act.[1] Despite this bland description, few issues created as much controversy during the RCRA reauthorization period as the issue of what to do about small generators. Although the final version of the small quantity generator provision contains fewer than a dozen paragraphs, the floor debates on the issue consumed more than 42 pages in the Congressional Record during the 98th Congress. The result of the extensive debate was a provision which will have widespread effect on the business community and, if enforced, create a major expansion of the EPA bureaucracy.

EPA estimates that the new 1984 requirements will increase the number of federally regulated generators from fewer than 15,000 to well over 100,000[2] with an incremental cost to industry of between $100 million and $300 million annually.[3] More than half of the small quantity generators fall into one of five industrial categories: auto body shops and other vehicle maintenance stations, manufacturing and finishing of metals, printing, photography, and laundries and dry cleaners.[4] (See Figure 1, next page.)

The small generator "loophole" originated on May 19, 1980, when EPA established a regulatory exclusion based on its own administrative convenience and resource constraints for those facilities generating less than 1,000 kilograms (one metric ton) of hazardous waste per month. In the preamble to those regulations, EPA acknowledged the need to regulate small quantity generators, but it said the effort required to do so would be beyond the agency's capabilities.[5] EPA did, however, commit itself to studying the problem of wastes from small quantity generators.

According to the House Energy and Commerce Committee, estimates of the amount of hazardous waste not regulated as a result of the small quantity generator exclusion ranged from 1 to 10 percent of the total amount generated.[6] While proponents of a lower threshold acknowledged that the amount in question did not represent a large percentage of the total, they principally were concerned that even a relatively small quantity of hazardous waste could be dangerous if handled improperly. For example, during the debate on H.R. 6307 (the House bill to reauthorize RCRA during the 97th Congress), Congressman John Sieberling (D-Ohio) recounted a particular incident in his own district:

> During the recess a small generator of toxic waste . . . who had gone bankrupt, sold some of the facilities that he had. The people who bought some of these facilities decided to open up a tank and find out what was in it. They turned one valve, and the fumes that came out nearly knocked them out. They managed to get away, but a toxic cloud of such magnitude ensued that the police had to evacuate 20 city blocks of a residential area. Afterward when the toxic gas cloud had been dissipated by the wind so the people could come back to their homes the authorities

Figure 1
Distribution of Small Quantity Generators by Industry Group:
SQGs and VSQGs

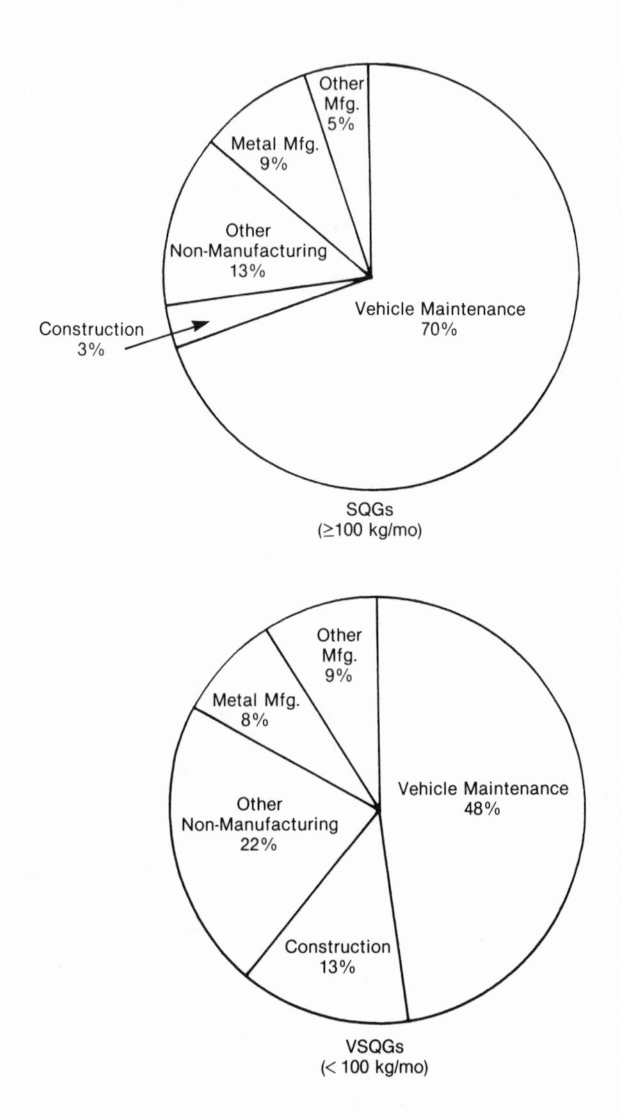

Source: Small Quantity Generator Survey data and analysis of secondary industries: 535,000 small quantity generators. This figure appeared in "National Small Quantity Hazardous Waste Generator Survey," February 1985, p.35.

> found almost a hundred barrels of various types of toxic waste that had been accumulated over a period of time and stored on this man's property.[7]

Another concern involved the realization that small quantity generators mixed their hazardous waste with conventional solid wastes, and that, as a result, unsuspecting trash collectors had been injured. Meanwhile the sanitary landfills which received the wastes became potential candidates for Superfund's National Priority List.[8]

Not all of the concerns, however, lay with those favoring more regulation of small quantity generators. The small business community was extremely agitated over the prospect that the paperwork and recordkeeping requirements would be overwhelming, and the compliance costs "excessive." There were real concerns, too, over the increased resource demands EPA and state agencies would face at a time of severe federal budget restraint.

After many months of debate by Congress and intense lobbying by the small business community, a consensus was reached and, on August 4, 1983, it was offered as an amendment to H.R. 2867 on the House floor by Congressmen Norman F. Lent, New York Republican, and Richard Shelby, Alabama Democrat. With relatively minor modifications, the "Shelby-Lent amendment" (which was also subsequently approved by the Senate) became the final version of the legislation.

Requirements for Small Quantity Generators (SQGs)

The 1984 Amendments require that by March 31, 1986, EPA promulgate standards for generators of waste in quantities of more than 100 and less than 1,000 kilograms per month. The standards "may vary" from those in effect for larger generators, but they must be sufficient to protect human health and the environment.[9] Although congressional intent was clearly to allow less stringent standards for the smaller generators, some questioned how less stringent standards could protect human health and the environment since presumably the large generator standards were also based on the same criteria.[10]

The Conference Report acknowledged the issue but concluded it was not a real problem. The conferees reasoned that in fulfilling its objective of protecting human health, EPA could choose among several approaches.[11] In making that choice, EPA was to take into account that many small quantity generators are small businesses that would be adversely affected if the full set of Subtitle C regulations were required.[12]

Additionally, Section 221 mandates that EPA must, at a minimum, require that all treatment and disposal of hazardous wastes from small quantity generators take place at permitted or interim status facilities.[13] Small quantity generators may, however, store hazardous wastes on site without a permit for up to 180 days, so long as the total quantity of hazardous wastes generated in a month is less than 1,000 kilograms. The storage period is extended to 270 days if the generator must ship the waste more than 200 miles and the total stored is less than 6,000 kilograms.[14]

The longer storage period (90 days is allowed for large quantity generators), according to the Senate Report, will allow the generator to consolidate wastes into larger loads for shipment off the premises. The period of 270 days for remote generators was chosen primarily to accommodate schools and universities, which could accumulate wastes through an academic year for a single shipment.[15]

Section 221 also provides EPA with explicit authority to regulate generators that produce *less* than 100 kilograms of hazardous wastes per month.[16] Although there was little if any doubt that EPA already had this authority, Congress wanted to underscore it since it was possible to construe the mandate to regulate down to 100 kilograms per month as an indication that going below the 100 kilograms threshold was unwarranted. Congress wanted to preempt that possible interpretation before it arose.

Interim Requirements

Until EPA promulgates the new small quantity generator regulations, small quantity generators may dispose of their wastes only at hazardous waste facilities with a Section 3005 permit issued by EPA, or at facilities permitted, licensed, or registered by a state to manage municipal or industrial waste.[17] Because this rule simply codifies existing regulatory requirements, Congress did not delay its effective date.[18] In its Codification Rule, EPA interprets this provision expansively in order to allow small quantity generator wastes to be disposed of at interim status facilities and those with RCRA permits issued by states with an approved hazardous waste program.[19] Moreover, EPA interprets Section 221 to allow the use of state-authorized municipal or industrial facilities for treatment, storage, and disposal, although the statutory provision seems to limit use to disposal facilities only.[20] Finally, although the statute does not specifically say so, EPA interprets the provision so as not to invalidate its existing regulation allowing for on-site accumulation of up to 1,000 kilograms of hazardous waste.[21]

Section 221 also requires that by August 9, 1985, a manifest must accompany all small quantity generator wastes shipped off-site. The manifest, although somewhat less burdensome than for large generators, must include: a) the generator's name and address; b) an accurate U.S. Department of Transportation description of the waste; c) the number and type of containers; d) the quantity of waste; and e) the name and address of the receiving facility.[22] In addition to extending the "cradle-to-grave" tracking system to small generators, the use of the manifest system addresses Congress' fundamental concern that notice be given to transporters, treaters, and disposers that the wastes they are handling are hazardous.[23] It is worth noting that the House bill would have required manifest notification by generators who produced as little as 25 kilograms per month.[24] In the Conference Committee, the chief advocate of the House position, Congressman James Florio (D-New Jersey), found little support for any threshold level lower than 100 kilograms per month, so the 25-kilogram threshold was abandoned.

In its Codification Rule, EPA points out that states with authorized programs may have additional manifest requirements applicable to small quantity generators. It therefore urges generators in those states to contact the appropriate state agency when completing the manifest form to ensure compliance with state law.[25]

The SQG Hammer Provisions

Many in Congress were well aware of EPA's reluctance to come to grips with the small quantity generator issue. They felt that, left to "agency discretion," the problem would only be studied interminably. Consequently, the 1984 Amendments contain a hammer provision for small quantity generators. If EPA fails to promulgate regulations on time (that is, by March 31, 1986), wastes produced by small quantity generators may be disposed of only at facilities with a permit (or interim status) and the following requirements will apply:

- the interim manifest requirements, including the name of the waste transporter;
- submission of the manifest exception reports on a semi-annual basis; and
- retention for three years of a copy of the manifest signed by the receiving facility.[26]

Reports to Congress

As in many other controversial matters affecting its small business constituents, Congress believed that launching studies of sensitive issues would somehow soften the regulatory burden. Since the small quantity generator issue was intensely controversial, no fewer than four studies were ordered. Consequently, EPA was directed to submit a comprehensive study by April 1, 1985, characterizing the generators, wastes, practices, and risks posed by wastes generated in quantities less than 1,000 kilograms per month.[27] This study, entitled "National Small Quantity Hazardous Waste Generator Survey," was issued in February 1985. Also, by April 1, 1985, EPA was directed to submit studies on a) the merits of retaining the existing manifest system for small quantity generator waste; b) the feasibility of establishing a licensing system whereby transporters assume the responsibilities for small quantity generators; and c) the problems associated with the disposal of hazardous wastes generated by educational institutions.[28] The deadlines for these studies were not met.

Informing Small Generators

During a 30-month period following enactment of the 1984 Amendments, EPA was directed by the Amendments to inform small quantity generators of their new statutory responsibilities.[29] Congress authorized $500,000 annually for fiscal years 1985 through 1987 to perform this task.[30] This amendment was offered by Congressman Ike Skelton (D-Missouri) in order to remedy the frequently cited problem of small quantity generators violating a law because they were unaware of its requirements.[31] "You have to publicize the requirements," Skelton declared. "You have to reach out and instruct small businesses on what is required. . . . [W]ith this money [EPA] could start a small program perhaps with seminars or newsletters . . . about what will be expected of them and how to comply with the law."[32] EPA has stated that it will work closely with state and national trade associations in order to spread the word to small quantity generators.

EPA's SQG Regulations

On August 1, 1985, EPA proposed regulations governing management, transportation, and disposal of hazardous wastes generated by small generators.[33] After making some revisions to its proposal, the final rule was promulgated on March 24, 1986, just six days before the hammer would have fallen. The regulations, which will affect generators producing between 100 kilograms and 1,000 kilograms per month, spell out a comprehensive regulatory scheme designed to be less burdensome than the full Subtitle C requirements.

Under the proposal, small quantity generators would be required to:

- Determine whether their wastes are hazardous (as is already required under 40 C.F.R. Section 261.5);
- Obtain an EPA Identification Number;
- Ship wastes only with a transporter having an EPA Identification Number;
- Use a multi-part round trip Uniform Hazardous Waste Manifest for off-site shipments;
- Maintain copies of manifests for three years;
- Comply with applicable Department of Transportation requirements for labeling and shipping wastes off-site; and
- Transport wastes only to approved Subtitle C facilities.[34]

With respect to on-site disposal, the regulation requires generators to comply with the interim status requirements under 40 C.F.R. Section 265 until the generator has obtained a Subtitle C permit. In addition, hazardous wastes stored in underground tanks for periods longer than the 180- or 270-day maximum would be subject to the proposed underground tank requirements, which will be imposed on all facilities using underground tanks.

It is worth emphasizing that the effect of the small quantity generator regulations will be to exempt small generators from many record-keeping requirements, including the biennial and "exception" reports.[35] The paperwork burden was one of the principal concerns of the small business community during the debate on this provision.

Figure 2
Breakdown of Hazardous Waste Generation
by Large and Small Quantity Generators

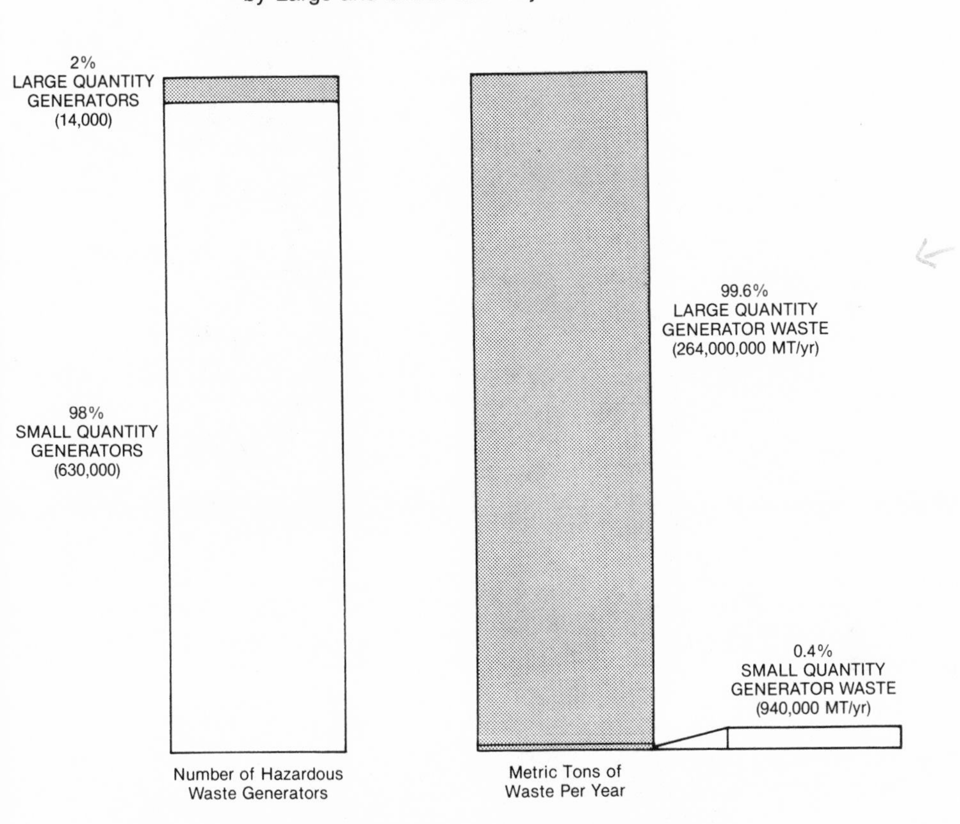

Source: Small Quantity Generator Survey data and analysis of secondary industries and additional incidental small quantity generators. Large quantity generator estimates based on *National Survey of Hazardous Waste Generators*, 1984. This figure appeared in "National Small Quantity Hazardous Waste Generator Survey," February 1985, p.29.

Figure 3
Breakdown of Small Quantity Generators
by Waste Quantity Category

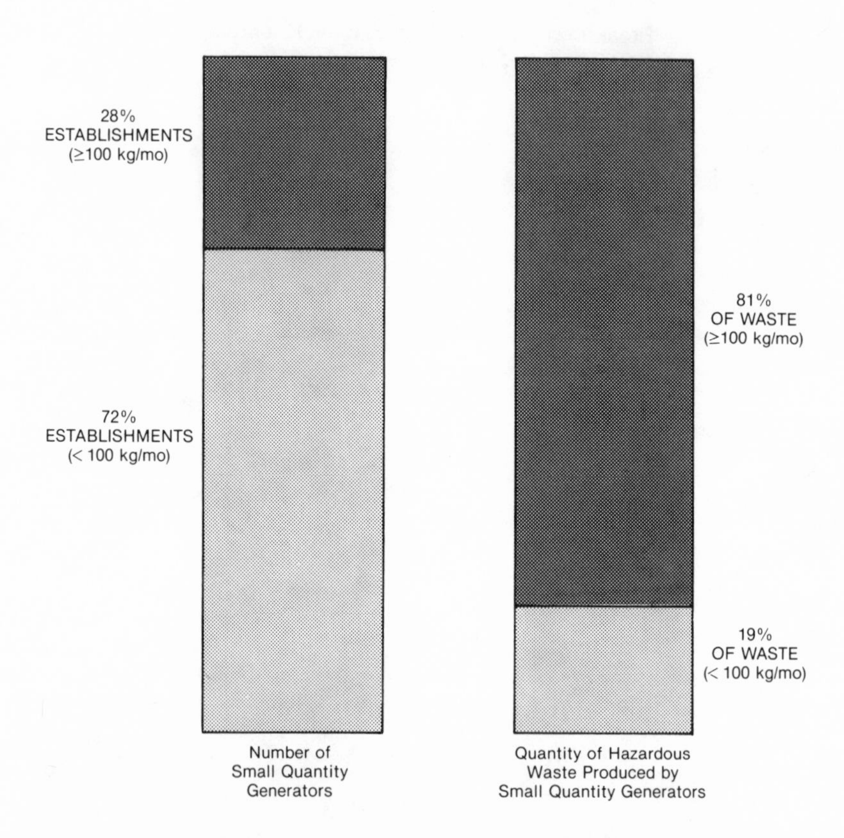

Source: Small Quantity Generator Survey data and analysis of secondary industries: 378,000 small quantity generators estimated from detailed information for targeted waste streams and 85,000 small quantity generators from secondary industries—691,000 metric tons of waste per year. This figure appeared in "National Small Quantity Hazardous Waste Generator Survey," February 1985, p.30.

Finally, small quantity generators would no longer be referred to, at least by EPA, by that term. That designation has been conferred, at least officially, on those who generate less than 100 kilograms of hazardous waste each month.[36]

EPA Study on Amount of SQG Wastes

During the debate on the small quantity generator provision, Congress was hindered by a lack of accurate data on the number of generators, how much waste they produce, and where it goes. Although EPA had been conducting a study of small quantity generators during the 98th Congress, the results of that study were not available until February 1985, about four months after the RCRA Amendments were enacted.

The study, which surveyed approximately 50,000 establishments, estimates that there are approximately 630,000 establishments that generate less than 1,000 kilograms per month of hazardous wastes, and produce a total of approximately 940,000 tons of hazardous waste annually. The approximately 455,000 generators of less than 100 kilograms per month account for only 180,000 tons of the hazardous waste generated annually. This leaves an estimated 175,000 generators of 100 to 1,000 kilograms per month, producing 760,000 tons of hazardous waste annually. That figure amounts to only 0.3 percent of the 264 million tons estimated to be generated annually in the U.S. (See Figures 2 and 3, from EPA study on previous two pages.)

Congress spent a disproportionate period of time debating the subject of small quantity generators when compared to the relative risk presented by these generators' wastes. That isn't to say that the risks posed are not substantial—940,000 tons of hazardous waste can, if improperly disposed of, cause widespread contamination, particularly since much of the waste consists of solvents and strongly acidic wastes. However, it is interesting to speculate on whether Congress—had the results of the study been available to it while it was considering the amendments—would have handled the subject of small quantity generators differently.

Chapter 10 Footnotes

[1] Senate Report at 7.

[2] EPA Brochure, *Small Quantity Hazardous Waste Generators: The New RCRA Requirements,* No. 530-SW-85-005 (March 1985); EPA Report, "Small Quantity Hazardous Waste Generator Survey," 31 (Feb. 1985).

[3] Congressional Budget Office, *Hazardous Waste Management: Recent Changes and Policy,* 53, Table 16 (May 1985).

[4] EPA Brochure, supra note 2.

[5] 45 Fed. Reg. 33103.

[6] House Report at 25.

[7] 128 Cong. Rec. H6764 (daily ed., Sept. 8, 1982).

[8] Senate Report at 7.

[9] HSWA Sec. 221(a), adding RCRA §3001(d)(2).

[10] Conference Report at 102.

[11] Conference Report at 103.

[12] Conference Report at 103; also House Report at 26.

[13] Adding RCRA §3001(d)(6).

[14] *Id.*

[15] Senate Report at 10.

[16] HSWA Sec. 221(a), adding RCRA §3001(d)(4).

[17] HSWA Sec. 221(a), adding RCRA §3005(d)(5).

[18] Senate Report at 9.

[19] Codification, 50 Fed. Reg. 28720 (July 15, 1985).

[20] Codification, 50 Fed. Reg. 28721.

[21] *Id.*, appearing in EPA regulations at 40 C.F.R. §261.5(f).

[22] Adding RCRA §3001(d)(3).

[23] Senate Report at 8; House Report at 27.

[24] House Report at 3 and 27.

[25] Codification, 50 Fed. Reg. 28720 (July 15, 1985).

[26] HSWA Sec. 221(a), adding RCRA §3001(d)(8).

[27] HSWA Sec. 221(c), citing no RCRA sections.

[28] HSWA Sec. 221(d), (e), and (f), citing no RCRA sections.

[29] HSWA Sec. 221(a), adding RCRA §3001(d)(9)(b).

[30] Conference Report at 3, 80.

[31] Senate Report at 27.

[32] 129 Cong. Rec. H6531 (daily ed., Aug. 4, 1983).

[33] 50 Fed. Reg. 31278.

[34] 51 Fed. Reg. 10146 (Mar. 24, 1986).

[35] 51 Fed. Reg. 10160 (Mar. 24, 1986). On June 6, 1986, the Environmental Defense Fund filed a petition challenging this component of the regulations with the U.S. Court of Appeals for the District of Columbia.

[36] 51 Fed. Reg. 10151 (Mar. 24, 1986).

Chapter 11

Burning and Blending

Summary

EPA Regulations for Burning Hazardous Wastes:

- Must be issued by November 8, 1986.
- Must be sufficient to protect human health and the environment.
- Exemption for certain petroleum refinery wastes converted to coke at same facility where generated.
- EPA may exempt burning of *de minimis* quantities under certain conditions.

Notification and Recordkeeping Requirements:

- By February 8, 1986, producers, burners, distributors, and marketers must notify EPA of operation.
- EPA Administrator may grant exemption from notification requirement.
- By February 8, 1986, EPA must issue recordkeeping requirements.

Labeling Requirements:

- Invoice or bill of sale must list hazardous wastes.
- Also must bear statement, "Warning: This Fuel Contains Hazardous Wastes."
- Not applicable to petroleum refining wastes from which all contaminants are removed and several other conditions met.

The "burning for energy recovery" exemption which EPA promulgated on May 19, 1980, allowed the unregulated burning of up to 20 million tons of hazardous waste each year.[1] Still in the early phases of promulgating hazardous waste regulations, the agency decided not to tackle this problem on the grounds that the burning of hazardous waste fuel provided a source of energy and accomplished at least some level of destruction.

The House Energy and Commerce Committee, however, saw things differently. According to the Committee, the practice "is one of several areas" in which EPA's decision not to regulate "has led to direct threats to human health and the environment."[2] The House Report also noted that as a result of this regulatory loophole, "[h]azardous wastes have been blended with heating oil and sold to unsuspecting customers."[3] Because approximately 98 percent of all boilers are small (less than 10 million BTU per hour capacity)[4] and therefore unable to achieve environmentally safe combustion and destruction efficiencies, the potential health risk from their dispersing large quantities of toxic constituents into the air is indeed significant.

By 1983, views on the practice of "burning for energy recovery" had shifted so much that closing the burning and blending regulatory loophole was one of the least controversial provisions of H.R. 6307, the RCRA reauthorization bill passed by the House during the 97th Congress. Even EPA, which had created the loophole, acknowledged that burning for energy recovery posed a serious risk to public health.[5]

An investigation by the staff of EPA's National Enforcement Investigations Center (NEIC) confirmed that the practice of mixing hazardous wastes into fuel oil was widespread and that it presented serious risks. According to the NEIC report:

> contaminants such as chlorinated solvents and polychlorinated byphenyls (PCBs) are added, either purposely or inadvertently, to the waste oil. The officials of several facilities in Ohio admitted to mixing solvents. . . . with the waste oil which is then sold as fuel. . . . [6]

The NEIC report also noted that "much of the waste oil in the New York City/New Jersey area is burned in lieu of #4 fuel oil in large residential and commercial buildings. . . . located in highly populated areas; hence, large numbers of people may be exposed to contaminants."[7]

Section 204 of the 1984 Amendments, which the Congressional Budget Office estimates will cost industry between $456 million and $1,620 million annually,[8] is designed to end the "burning for energy recovery" exemption by requiring EPA to regulate generators, transporters, distributors, marketers, and burners of fuel derived from or blended with hazardous waste. The regulations, which EPA was directed to promulgate by November 8, 1986, must be sufficient to protect human health and the environment.[9]

A statutory exemption is provided for petroleum refinery wastes which contain oil and which are converted to petroleum coke at the same facility where the wastes are generated. This exemption does not apply if the resulting coke product fails one or more of the characteristic tests (toxicity, ignitability, corrosivity or reactivity) established under Section 3001.[10]

EPA is authorized to exempt facilities that burn *de minimis* quantities of hazardous waste as a fuel so long as: (1) the wastes are burned at the facility at which they are generated; (2) the waste is burned to recover useful energy; and (3) the waste is burned in a device determined by the EPA Administrator to be designed and operated in a way that protects human health and the environment.[11]

Until applicable standards are issued, no hazardous waste fuel (any commercial product which is burned or processed as a fuel rather than used for its originally-intended purpose) may be burned in any cement kiln located in a municipality with a population of more than 500,000, unless the kiln complies with RCRA incinerator requirements.*[12] This provision resulted from the prospect that a particular cement kiln in Congressman Martin Frost's district in Dallas, Texas, would use hazardous wastes as fuel.

A key element of Section 204 is the notification provision which requires that, by February 8, 1986, producers, burners, distributors, and marketers of hazardous waste fuel or used oil fuel** provide EPA or an authorized state with the following information:

- the location and general description of the facility;
- a description of the hazardous waste;
- a description of the production or energy recovery activity carried out at the facility.[13]

Notification is not required in specific cases for which the Administrator specifically determines is unnecessary. Also by February 8, 1986, EPA must issue regulations requiring that any person required to file notification maintain appropriate records.[14]

Section 204 also establishes certain labeling requirements.[15] Until the Administrator promulgates superseding regulations, it is unlawful to distribute, even within a company,[16] or market any hazardous waste fuel unless the invoice or bill of sale bears the statement "Warning: This Fuel Contains Hazardous Wastes," and lists the hazardous wastes contained in the fuel. EPA says in its Codification Rule that this listing requirement may be satisfied by generic rather than precise chemical identification.[17]

In addition, unless the Administrator determines otherwise, the labeling requirement does not apply to fuels produced from petroleum refining waste containing oil and fuels produced from oily materials resulting from normal petroleum refining, production, and transportation practices, provided several conditions are met, including the removal of all contaminants.[18]

*In its Codification, EPA stated that this provision does not apply to cement kilns burning hazardous wastes for material recovery. Where the burning is for both energy and material recovery, EPA says, the provision is applicable. 50 Fed. Reg. 28724 (July 15, 1985).

*The 1984 Amendments authorize EPA to regulate used oil as a hazardous waste, as discussed below in Chapter 15 on Solid and Oil Waste Minimization. EPA has proposed rules for burning used oil which differ significantly from regulations governing burning of hazardous waste.

On January 11, 1985, EPA proposed regulations governing the burning of hazardous waste fuels and of fuels derived from used oil and on November 29, EPA issued a final rule.[19] In essence, EPA's rule would create a clear regulatory distinction between used oil which is acceptable for burning in nonindustrial boilers, and that which is not. Fuel that meets EPA's stringent criteria would be virtually exempt from federal regulation, including from requirements for notification or tracking. (The fuel producer would, however, have to document through analyses that the oil meets the specifications, and would also have to comply with certain recordkeeping requirements.) According to the agency, burning fuel that satisfies the "specification" criteria (see chart) would not pose hazards significantly greater than those resulting from the burning of virgin oil.[20]

Used Oil Fuel Specification

Constituent/ property	Proposal allowable level	Final rule allowable level
Arsenic	5 ppm maximum	5 ppm maximum
Cadmium	2 ppm maximum	2 ppm maximum
Chromium	10 ppm maximum	10 ppm maximum
Lead	10-100 ppm maximum	100 ppm maximum
PCBs	50 ppm maximum	
Total halogens,		4000 ppm maximum
Flash point	100 °F minimum	100 °F minimum

In addition to the fuel specifications, EPA will also require marketers, distributors, and burners of hazardous waste fuel and "off-specification" used oil fuel to notify EPA of their "waste-as-fuel" activities. After notification, the agency will assign each marketer and burner an EPA Identification Number.[21]

The notification requirements are designed to provide EPA with the number and location of facilities involved in processing, blending, marketing and distributing of waste fuels as well as the number, type and location of burners. Notification is also a prerequisite for obtaining interim status for hazardous waste storage facilities. The agency has estimated that between 20,000 and 30,000 persons will be required to file notification.[22]

EPA's final rule also prohibits the burning of hazardous waste and off-specification used oil fuel in non-industrial boilers—such as those used to heat apartment and office buildings, schools and hospitals. EPA's rule allows continued burning of hazardous waste and off-specification used oil fuel by industrial and utility boilers since the destruction efficiencies of these boilers are capable of preventing serious health risks. Moreover, many industrial facilities are equipped with air pollution devices which are capable of preventing the emission of toxic constituents. However, EPA has indicated that it plans to propose additional restrictions in the form of technical and permit standards in 1986.[23]

The final rule also requires that:

- all shipments of hazardous waste fuel be accompanied by a manifest. Marketers of off-specification used oil fuel shipments send an invoice to the buyer. However, the invoice may be sent directly; it need not accompany the shipment.
- marketers and buyers must keep copies of the manifests or invoices for a period of three years from shipment;
- all storage of hazardous waste fuel must comply with the applicable standards set forth in 40 C.F.R. Parts 262, 264, 265, and 270.
- transporters of hazardous waste fuel must comply with the basic transportation requirements set forth in 40 C.F.R. Part 263.

Chapter 11 Footnotes

[1] Senate Report at 36.
[2] House Report at 39.
[3] *Id.*
[4] 50 Fed. Reg. 1687 (Jan. 11, 1985).
[5] 48 Fed. Reg. 14481-82 (April 4, 1983).
[6] National Enforcement Investigation Center, Summary of Waste Oil Recycling Investigations, 4 (October 1983).
[7] *Id.* at 5.
[8] Congressional Budget Office, *Hazardous Waste Management: Recent Changes and Policy Alternatives,* 53, Table 16 (May 1985).
[9] HSWA Sec. 204(b), adding RCRA §3004(q)(1).
[10] HSWA Sec. 204(b), adding RCRA §3004(q)(2)(A).
[11] HSWA Sec. 204(b), adding RCRA §3003(q)(2)(B).
[12] HSWA Sec. 204(b), adding RCRA §3004(q)(2)(C). This requirement appears in EPA regulations at 40 C.F.R. §266.31.
[13] HSWA Sec. 204(a), amending RCRA §3010.
[14] HSWA Sec. 204(b), adding RCRA §3004(s).
[15] HSWA Sec. 204(b), adding RCRA §3004(r)(1).
[16] Codification, 50 Fed. Reg. 28724 (July 15, 1985).
[17] *Id.*
[18] HSWA Sec. 204(b), adding RCRA §3004(r)(2) and (3).
[19] 50 Fed. Reg. 1683 (proposed rule); 50 Fed. Reg. 49164 (final rule).
[20] 50 Fed. Reg. 1695.
[21] 50 Fed. Reg. 1702-1703.
[22] 50 Fed. Reg. 49196 (November 29, 1985).
[23] 50 Fed. Reg. 49192 (November 29, 1985).

Chapter 12

Special Wastes

Summary

Mining, Utility, and Cement Kiln Wastes:

- Authority to modify technical requirements for treatment, storage, and disposal if listed as hazardous wastes.
- Requirements must reflect practical difficulties.

Uranium Mill Tailings:

- RCRA Amendments do not affect the Uranium Mill Tailings Radiation Control Act.

Domestic Sewage Exemption for Hazardous Waste Disposal:

- By February 8, 1986, EPA must submit a report on hazardous waste excluded by this exemption.
- Eighteen months later, EPA must issue regulations to control these hazardous wastes.
- By November 8, 1987, EPA must submit a report on the groundwater effects of wastewater lagoons at sewage plants.

Mining, Utility and Cement Kiln Wastes

Since enactment of the "Bevill Amendment" in 1980 (named for its author Congressman Tom Bevill of Alabama) Congress has given certain waste streams special consideration. The principal impetus for this special treatment was the problem presented by the prospect of regulating mining wastes.

These wastes are large in quantity, but relatively low in toxicity. The estimated 1.75 billion tons of mining wastes generated each year exceeds the total volume of solid waste generated by all other industries combined.[1] Wastes generated by the utility industry and cement kiln dust wastes were considered to have similar "high volume/low toxicity" characteristics that might warrant something other than the generally-applicable technological requirements. Because each of these "special" wastes had been precluded from regulation by the 1980 Amendments,[2] and because EPA had not completed its studies of these wastes, Congress in 1984 simply decided to continue the policy of dealing with the wastes separately.

In order to take into account the "special characteristics" of these wastes, the 1984 Amendments authorize EPA to modify the technical requirements otherwise applicable to treatment, storage and disposal practices.[3] It was generally agreed, for example, that maintaining the integrity of a liner under the weight of massive mining wastes would be extremely difficult, if not impossible, and the public health arguments for making the effort were not convincing. Thus, Section 209 provides that if EPA decides to list any of these wastes as hazardous, it must consider the practical difficulties of implementing regulatory controls.* Although Section 209 makes clear that the modified requirements for these wastes must be sufficient to protect human health and the environment, according to the Conference Report, this standard "does not necessarily imply the uniform application of requirements developed for disposal of other hazardous wastes."[4] The Senate Report notes that the provision "does not preclude EPA from requiring double lining of landfills or surface impoundments for mining and mineral processing wastes in those cases where it is appropriate to do so."[5]

On June 30, 1986, EPA stated its intention to regulate mining wastes, i.e., waste from the extraction and benefication of ores and minerals, under Subtitle D rather than Subtitle C. The Agency had concluded that the Subtitle C standards would be "environmentally unnecessary, technically infeasible or economically impractical" when applied to mining waste. However, EPA also made clear that if it was unable to develop an effective mining waste program under Subtitle D, it would consider using its authority under Subtitle C.

Uranium Mill Tailings

One of the issues which for several months delayed the Senate's considera-

*On December 31, 1985, EPA issued a report on the quantity of wastes generated by certain segments of the mining industry and the nature of the hazards posed by these waste streams.

tion of S.757 involved uranium mill tailings—the large volumes of sand-like material that result from the processing of uranium ore. On June 20, 1984, more than one year after the Senate Environment and Public Works Committee had approved a RCRA reauthorization bill, Senators Pete Domenici (R-New Mexico), Alan Simpson (R-Wyoming), and nine other Republican senators from western states wrote a letter to Senate Majority Leader Howard Baker (R-Tennessee), making clear their position that a long-standing dispute about the applicability of RCRA to uranium mill tailings needed to be resolved. In that letter, the senators, most of whom had placed a "hold" on S.757 to prevent its consideration, stated that "upon resolution of this issue, we trust that this one major remaining impediment to bringing this bill up on the Senate floor will have been removed." Resolving the issue of how to deal with uranium mill tailings *before* taking the bill to the floor, the senators told Baker, would "avoid the need for an extended discussion of this complex issue on the Senate floor whereby we would hope to educate those members who may not be as familiar with this matter as others"— only a slightly veiled threat to engage in a filibuster.

What was this "complex issue" which for a time threatened the reauthorization of RCRA? Curiously, the issue hardly created any discussion during the deliberations by the Environment and Public Works Committee on S.757. But it was not a new controversy. In 1978 Congress in the Uranium Mill Tailings Radiation Control Act (UMTRCA)[6] directed EPA to promulgate regulations to protect public health and the environment from radiological and nonradiological hazards created by uranium mill tailings. The standards to control nonradiological hazards were to be "consistent with the Solid Waste Disposal Act." Once EPA had come up with its standards, the Nuclear Regulatory Commission was supposed to incorporate the standards into the general environmental requirements uranium mills would have to meet.

When EPA issued its regulations on October 7, 1983,[7] the uranium industry was critical, particularly with respect to the requirements for liners to prevent groundwater contamination. EPA said it had adopted the liner requirement because it was central to the RCRA system and therefore necessary in order to comply with the provision in UMTRCA that mandated consistency with RCRA.

The uranium industry argued that EPA's regulations were not supposed to specify what the industry considered "detailed engineering methods," which from the industry's standpoint included any requirements for liners. According to the industry, EPA's role was limited to establishing "general standards of performance."

However, the industry also objected to the clearly "general standards of performance" in EPA's October 7 regulations. According to the industry, by establishing concentration limits of contaminants (which were to be measured 500 feet from a tailings pile), EPA had impermissibly interfered with NRC's jurisdiction because any activity within the boundaries of the mill was, or should be, off-limits to EPA.

To address the difficulties for the uranium industry created by EPA's regulations, and to prevent similar problems from arising in the future, Senator Simpson proposed two amendments. The first, intended to force EPA to revise its October 7 regulations, included prohibitions on "specific management requirements" and prohibitions on standards that "are intended to be site specific in nature." In addition, the proposed amendment would have required that "such standards shall not impose requirements applicable within the boundaries of uranium milling facilities."

The second amendment was designed to prevent EPA from imposing any new standard that would become applicable as a result of the 1984 RCRA reauthorization. EPA would be precluded from imposing such additional burdens as double-liners, leachate collection systems, and leak detection equipment.

After five weeks of negotiations with Senators Mitchell, Chafee, Randolph, Stafford, and the Senate leadership, Senator Simpson agreed to release his hold of S.757 in return for adoption of an amendment that would simply assure that the 1984 Amendments "do not affect, modify, or amend" UMTRCA. This amendment was adopted by the Senate when it passed S.757 on July 25, 1984.

The Conference Committee adopted the Simpson amendment without any modification. However, the legislative history in the Conference Report, which was agreed to after considerable discussion, was intended to short circuit any attempt to read into the amendment more meaning than what was contained in the actual statutory language. In its entirety, the Conference Report's illumination of the Simpson amendment reads as follows: "Nothing in this section shall be deemed to *preclude or to require* the revision of existing regulations promulgated under UMTRCA." [Emphasis added.][8] Whether this ambiguous compromise ends the confusion and controversy surrounding the relationship between UMTRCA and RCRA remains to be seen.

Closing the 'Domestic Sewage' Loophole

Perhaps the largest loophole in the RCRA program as it existed in the early 1980s can be found in the statute itself, and therefore, unlike the small quantity generator loophole, cannot be blamed on EPA's implementation of Subtitle C.

In 1976, Congress had defined the term "solid waste" to include virtually any discarded material, but not "solid or dissolved material in domestic sewage . . . "[9] During the floor debate on H.R. 2867, Congressman Guy Molinari (R-New York) complained that: "[t]his unqualified exclusion invites abuse. Generators can use the cheapest method to dispose of their wastes—simply dumping it down the sewer."[10] Molinari's concern was apparently fully justified. According to a recent survey by the Chemical Manufacturers Association, the exemption has allowed approximately 50 million tons of hazardous waste to go unregulated each year.

Instead of removing the seven words creating the loophole, however, Congress required that by February 8, 1986, EPA submit a report on hazardous wastes which are currently exempt from regulation as hazardous because they are mixed with domestic sewage or other wastes that pass through a sewer system to a publicly owned treatment works (POTWs).[11] Then, within 18 months of issuing the report, EPA must adopt regulations to assure that these wastes are adequately controlled to protect human health and the environment.[12] Additionally, EPA must submit a report to Congress by November 8, 1987, on wastewater lagoons at POTWs and their effect on groundwater.[13] The "study now, regulate later" approach in dealing with the domestic sewage loophole reflects, among other things, Congress' appreciation of the complexity of combining the authorities of the Clean Water Act, which regulates POTWs, with those of RCRA.

To help EPA gather adequate data for its domestic sewage studies, Congress provided that the authority to conduct inspections (Section 3007) and require notification (Section 3010) would extend to solid or dissolved materials in domestic sewage just as they apply to other hazardous wastes.[14]

On February 7, 1986, EPA submitted its report to Congress. The Agency stated that while hazardous waste discharged into sewers caused a wide range of environmental problems, including contamination of groundwater from leaking pipes and sewage spread on land, it could not quantitatively define the environmental effects or formulate a comprehensive regulatory strategy to control the environmental damage.

The report, entitled "Report to Congress on the Discharge of Hazardous Wastes to Publicly Owned Treatment Works," estimated that 178,000 metric tons of wastes considered hazardous under RCRA are discharged annually into sewer systems. Of that figure, approximately 92,000 tons would also be considered hazardous substances under the Clean Water Act. The report also stated 62,000 tons consisted of hazardous heavy metals and cyanide and up to 52,000 tons are organic chemicals.

The report recommended improvements in both federal and categorical pretreatment standards and local pretreatment regulations (particularly with respect to solvents) and more aggressive enforcement of existing Clean Water Act regulations. In addition, the report suggested that additional research was necessary and that a coordinated regulatory approach using RCRA, the Clean Water Act, and Superfund should be considered.

Chapter 12 Footnotes

[1] Senate Report at 28.
[2] *See* RCRA §§3001(b)(3)(A) and 8002(f), (n), (o), and (p).
[3] HSWA Sec. 209, adding RCRA §3004(x).
[4] Conference Report at 93.
[5] Senate Report at 29.
[6] P.L. 95-604, 42 U.S.C. §2022.
[7] 48 Fed. Reg. 45946; 40 C.F.R. §192, Subparts D and E.
[8] Conference Report at 129.
[9] RCRA §1004(27).
[10] 129 Cong. Rec. H9149, (daily ed., Nov. 3, 1983).
[11] HSWA Sec. 246(a), adding RCRA §3018(a).
[12] HSWA Sec. 246(a), adding RCRA §3018(b).
[13] HSWA Sec. 246(a), adding RCRA §3018(c).
[14] HSWA Sec. 246(a), adding RCRA §3018(d).

Chapter 13

Listing and Delisting Requirements

Summary

EPA Must List as Hazardous Wastes Where Appropriate:

- By May 8, 1985, chlorinated dioxins and chlorinated dibenzofurans.
- By November 8, 1985, remaining halogenated dioxins and halogenated dibenzofurans.
- By February 8, 1986, a number of specified wastes including refining wastes, solvents, inorganic chemical industry wastes, and paint production wastes.

EPA Must Revise Listing Criteria:

- By November 8, 1986, identify additional characteristics of hazardous substances, including indicators of toxicity.
- By March 8, 1987, make changes in the extraction process for determining toxicity.
- Identify hazardous wastes that are hazardous solely because they contain recognized hazardous constituents (e.g., known carcinogens).

In Delisting Hazardous Wastes, EPA Must:

- Consider factors in addition to those for which the waste was originally listed.
- Decide current temporary delistings by November 8, 1986.
- Decide new petitions for delisting within two years of the application to the maximum extent practicable.

When David Lennett of the Environmental Defense Fund (EDF) testified before the House Subcommittee on Commerce, Transportation and Tourism on March 24, 1983, he faulted EPA for not listing new hazardous wastes and for its hurry in granting "temporary delistings."* Lennett contended that the "temporary delisting" was a particularly insidious practice because it allowed the agency to cease regulating a waste "without examining all the constituents of concern in the waste . . . and without notice and an opportunity for public comment."[1] Equally disturbing, according to Lennett, EPA's "decisions [on delisting for toxicity] are based on a technically inadequate chemical procedure." The Extraction Procedure (EP) test for toxicity was viewed as unreliable for several reasons. For example, the EP test cannot be used to analyze organic wastes, and it is not an accurate predictor of the concentrations of heavy metals that leach from landfills.[2] Largely in response to the concerns raised by EDF and others in the environmental community, Congress required that EPA do a better job in listing and delisting hazardous wastes.

Listing Specific New Wastes

The 1984 Amendments require that by May 8, 1985, EPA, where appropriate, must list as hazardous waste those wastes containing chlorinated dioxins and chlorinated dibenzofurans. EPA issued the dioxin listing regulation on January 14, 1985 (50 Federal Register 1978). By November 8, 1985, EPA, where appropriate, must list as hazardous wastes those wastes containing halogenated dioxins and halogenated dibenzofurans.**[3] Previously, EPA had listed only one out of approximately 75 different forms of chlorinated dioxin as a hazardous waste.

By February 8, 1986, EPA must determine whether to list as hazardous the following wastes: chlorinated aliphatics, dioxin, dimethyl hydrazine, TDI (toluene diisocyanate), carbamates, bromacil, linuron, organo-bromines, solvents, refining wastes, chlorinated aromatics, dyes and pigments, inorganic chemical industry wastes, lithium batteries, coke byproducts, paint production wastes, and coal slurry pipeline effluent.[4] With the exception of the coal slurry pipeline effluent, EPA already was preparing to list these various wastes as hazardous. (The requirement to list coal slurry pipeline effluent as a haz-

*Wastes are regulated as hazardous under RCRA either because they or the industry process are specifically listed or they meet criteria established by EPA. RCRA Section 3001(a) and (b). Listed wastes are set forth at 40 C.F.R. §261.31-33, and characteristic or identified wastes at 40 C.F.R. §261.21-24.

**In a letter of August 19, 1985, to Senator Robert T. Stafford, the EPA Assistant Administrator for Solid Waste and Emergency Response, J. Winston Porter, stated that EPA would not be able to meet the November 8, 1985, deadline for listing halogenated dioxins and dibenzofurans. Among other recent EPA listing decisions were a proposal of April 30, 1985, to list mixtures containing spent solvents as opposed to just individual solvents, 27 of which are now listed. 50 Fed. Reg. 18378. The proposed rule would regulate mixtures containing a total of 10 percent or more by volume of one or more listed solvents. Also on July 30, 1985, EPA proposed to add benzene and three other organic solvents to the list of hazardous wastes. 50 Fed. Reg. 30908.

ardous waste had far more to do with the railroad lobby's efforts to slow the development of coal slurry pipelines than with the effluent's hazardous properties *per se.*) But, as the Senate Report noted, "EPA's listing process has been virtually stalled for several years."[5] Thus, one of the purposes of the statutory provision was to give EPA a congressional deadline on listing specific wastes.

Although polychlorinated biphenyls (PCBs) are not among the wastes to be reviewed for listing, the Conference Report makes it clear that EPA should proceed with its plans to list PCBs "as expeditiously as possible."[6] In addition, the 1984 Amendments contemplate the listing of PCBs by providing that, if a facility has received a permit for the incineration of PCBs under Section 6(e) of the Toxic Substances Control Act, it would be deemed to have a permit under RCRA.[7]

New Listing Criteria

In order to improve the process of regulating more hazardous wastes, Congress required EPA to modify three aspects of its listing criteria. First, by November 8, 1986, EPA must issue regulations identifying additional characteristics of hazardous substances, including indicators of toxicity.[8] The Senate Report named as a serious deficiency in EPA's listing procedures its failure to include "any characteristic which identifies wastes that pose a problem due to toxic organic constituents."[9] Second, by March 8, 1987, EPA must make changes as necessary to the extraction procedure (the "EP test") for determining toxicity, including changing the leaching media to ensure that it accurately predicts the leaching potential of the waste.[10] Finally, in cooperation with the Agency for Toxic Substances and Disease Registry and the National Toxicology Program, EPA must identify or list hazardous wastes that are hazardous solely because they contain recognized hazardous constituents (e.g., known carcinogens) at levels which endanger human health.[11]

New Delisting Criteria

Previously, for a waste to be delisted, it was necessary to demonstrate only that the waste did not meet any of the criteria for which it had been listed as hazardous.[12] Thus, according to the Senate Report, an applicant's waste could be considered nonhazardous with respect to the listed constituents and exempted from regulation, yet still be hazardous because of constituents that were not considered in the original listing determination.[13]

Now under the 1984 Amendments, before EPA can delist a waste, it must consider factors, including additional constituents other than those for which the waste was originally listed, that could cause the waste to be listed as hazardous.[14] Thus, for example, a petition to delist a wastewater treatment sludge from electroplating operations would have to show not only that the concentration for which the waste was listed is below levels of regulatory concern, but also that no other heavy metals or other constituents are present that would cause the waste to be hazardous.

Deadlines for Deciding Delisting Petitions

Prior to the 1984 Amendments, there was no statutory deadline by which EPA had to make a decision on a delisting petition. EPA issued temporary delistings without notice and comment and with no obligation to make a final decision.[15] Section 222 of the 1984 Amendments requires that EPA make a final decision by November 8, 1986, on temporary delistings issued before enactment of the Amendments. Temporary delistings not ruled on by that deadline expire.[16] This provision also requires that EPA, to the maximum extent practicable, make decisions regarding new petitions within two years of the date it receives a complete application. It should be noted that there is no counterpart to the self-executing expiration of previously granted temporary delistings; in other words, petitions for delisting are not granted by operation of the statute if the agency does not act within the specified time limits. Finally, given that dissatisfaction with the lack of notice and comment for temporary exclusions was a major factor leading to the revision of the delisting procedure, it is not surprising that the agency now is required to provide an opportunity for notice and comment on both temporary and permanent delisting petitions.[17]

Chapter 13 Footnotes

[1] Hearings before the House Subcommittee on Commerce, Transportation, and Tourism, Report No. 98-32, 98th. Cong., 1st. Sess. 435 (March 24, 1983).

[2] *Id.* at 472.

[3] HSWA Sec. 222(a), adding RCRA §3001(e)(1).

[4] HSWA Sec. 222(a), adding RCRA §3001(e)(2).

[5] Senate Report at 4.

[6] Conference Report at 105.

[7] HSWA Sec. 211(3), amending RCRA §3005(a).

[8] HSWA Sec. 222(a), adding RCRA §3001(h).

[9] Senate Report at 34.

[10] HSWA Sec. 222(a), amending RCRA §3001(g). On June 13, 1986, EPA proposed a rule that would expand the EP toxicity test to include 38 additional organic chemicals including arsenic, benzene, mercury, vinyl chloride, carbon tetrachloride, methylene chloride, toluene, methyl ethyl ketone, and various heavy metals. The rule would also improve the methodology for testing contaminants that leach into groundwater. One result of the new characteristic test will be to increase by 3 million tons annually the amount of waste considered hazardous. 51 Fed. Reg. 21648 (June 13, 1986).

[11] HSWA Sec. 222(b), amending RCRA §3001(b)(1).

[12] 40 C.F.R. §260.22.

[13] Senate Report at 33.

[14] HSWA Sec. 222(a), adding RCRA §3001(f)(1). The new delisting criteria are reflected in EPA regulations at 40 C.F.R. §260.22.

[15] Senate Report at 33.

[16] Adding RCRA §3001(f)(2)(B).

[17] Adding RCRA §3001(f)(1); see Codification, 40 Fed. Reg. 28727-28 (July 15, 1985).

Chapter 14

Hazardous Waste Minimization

Summary

- After September 1, 1985, large quantity generators must certify in their manifests that the volume or quantity and toxicity of the hazardous wastes has been reduced.
- After September 1, 1985, large quantity generators must certify the same information as a condition for an on-site permit.
- Large quantity generators must specify in biennial reports their waste reduction efforts.
- By October 1, 1986, EPA must submit a report on the utility of establishing specific requirements for reducing volume and toxicity of hazardous wastes.

Senator Robert Stafford of Vermont, chairman of the Environment and Public Works Committee, was not comfortable with the idea that America's hazardous waste problems could be solved by building better landfills. During the debate on S.757, he repeatedly emphasized the importance of reducing not only the management risks of hazardous wastes, but also the quantities produced (see Table 8, next page):

> I believe we are making some headway in protecting this Nation from the harm caused by hazardous wastes. But the fact remains that neither existing law nor S. 757 guarantees anything like truly safe disposal of the vast quantities of hazardous waste generated in this Nation.
>
> We need to find ways to greatly reduce the amount of harmful wastes. At the Federal level we must look for ways to encourage the development of alternative production processes, the substitution of less hazardous materials for more toxic ones, and other means to reduce both the quantity and toxicity of wastes.
>
> We must also provide incentives for development of treatment and disposal practices that minimize threats from those hazardous wastes that continue to be produced.[1]

Echoing Senator Stafford's concern, the Senate Report pointed out that "[r]egardless of the care with which [hazardous waste] facilities are managed and the regulatory or legal responsibilities imposed on these facilities, assuring protection of public health and the environment long after the active phase of a facility's existence has ended is a difficult task."[2]

In response to Senator Stafford's insistence that hazardous waste be minimized and not just regulated more stringently, Congress adopted a four-part program. The waste minimization program, outlined below, does not mandate particular changes in the production process, but rather is intended to encourage individual generators to make changes appropriate to their specific operations.[3] Senator Stafford called this program "a starting point, not the last word," in reducing the amount of harmful wastes.[4]

Manifests

Section 224 requires that after September 1, 1985, generators certify in their manifests that the volume or quantity and toxicity of the hazardous wastes they produce have been reduced to the degree economically practicable.[5] According to the Senate Report, "the determination of 'economically practicable' will be made by the generator and is not subject to subsequent re-evaluation."[6]

Generators also must certify in their manifests that the proposed method used to manage the waste minimizes threats to health and the environment to the extent practicable.[7] These requirements are contained in the revised Uniform Hazardous Waste Manifest Form.[8] Although EPA originally considered small quantity generators to be exempt from the certification require-

Table 8

Estimated Changes in Waste Generation Patterns, 1983 and 1990, By Major Industry Group Under Alternative Cases (In thousands of metric tons)

Major Industry	Quantity in 1983[a]	1990 Quantity With No Waste Reduction[b]	1990 Quantity With Waste Reduction[b]	Percent Change in 1990 from Waste Reduction
Chemicals and Allied Products	127,245	136,678	115,167	−9
Primary Metals	47,704	49,597	41,611	−13
Petroleum and Coal Products	31,358	29,213	25,526	−19
Fabricated Metal Products	25,364	25,493	11,820	−54
Rubber and Plastic Products	14,600	17,954	17,252	+18
Miscellaneous Manufacturing	5,614	5,856	5,001	−11
Nonelectrical Machinery	4,859	5,717	4,831	−1
Transportation Equipment	2,977	3,243	2,781	−7
Motor Freight Transportation	2,160	2,160	1,836	−15
Electrical and Electronic Machinery	1,929	2,313	1,557	−19
Wood Preserving	1,739	2,095	1,743	0
Drum Reconditioners	45	45	16	−64
Total	265,595	280,364	229,141	
Percent change from 1983	—	+5.6	−13.7	

SOURCE: Congressional Budget Office, *Hazardous Waste Management: Recent Changes and Policy Alternatives,* May 1985.

a. Mean estimate of waste generation under pre-1984 RCRA policies (see Table 3 in Chapter II of CBO Report).

b. Assumes no waste reduction efforts by industry in response to 1984 RCRA Amendments. Projection of waste generation levels in 1990 based on growth in industrial output levels forecast by the CBO generation model. Decreases in waste quantities therefore result from declining levels of industrial (and waste-producing) activity.

c. Assumes waste-reduction efforts by industry, as detailed in CBO Report, Table 11. Forecast includes waste growth from increases in industrial output identical to that of the higher 1990 waste-reduction case.

ment, the Agency has subsequently reassessed its position and has solicited comments on the applicability of the requirements to generators in the 100 to 1,000 kilograms per month category.[9]

Permits

After September 1, 1985, as a condition for an on-site permit, a generator must certify, at least annually, the same information described in the "Manifests" section above.[10] The Senate Report makes clear, however, that this requirement does not authorize EPA or anyone else "to interfere with or intrude into the production process or production decisions of individual generators."[11] Thus, the generator's failure to reduce the quantity of his waste stream could not be the basis for a citizen suit under Section 7002.

Generator Reports

Generators must specify, in biennial reports, their efforts to reduce waste volume and the reduction actually obtained.[12] According to the Senate Report, EPA, however, "should not require reports that duplicate the agency's existing biennial reports."[13]

EPA Study

Another amendment introduced by Senator Stafford requires that by October 1, 1986, EPA submit a report to Congress on the feasibility and desirability of expanding Subtitle C to include requirements for generators to reduce the volume and toxicity of their wastes.[14] According to the Senate Report, one type of requirement "which the Agency should evaluate . . . would be substantive standards of performance."[15] Senator Stafford said he expected that this study, along with information gathered from the biennial generator reports, would provide the basis for sound future changes that emphasize waste minimization.[16]

Chapter 14 Footnotes

[1] 130 Cong. Rec. S9147 and S9148 (daily ed., July 25, 1984).

[2] Senate Report at 65.

[3] Senate Report at 6 and 66.

[4] 130 Cong. Rec. S9147-48 (daily ed., July 25, 1984).

[5] Adding RCRA §3002(b)(1). On March 24, 1986, EPA proposed regulations concerning the waste minimization requirements imposed by this provision. According to EPA, the manifest certification aspect of Section 3002(b) "merely encourages generators to consider alternative waste management practices that would be more environmentally protective and more economically advantageous from the generator's perspective . . ." 51 Fed. Reg. 10178 (Mar. 24, 1986).

[6] Senate Report at 66.

[7] HSWA Sec. 224(b), adding RCRA §3002(b)(2).

[8] This form is published as an appendix to 40 C.F.R. Part 262 (Statement 16).

[9] 51 Fed. Reg. 10177 (Mar. 24, 1986).

[10] HSWA Sec. 224(b), adding RCRA §3005(h); reflected at 40 C.F.R. §262.41.

[11] Senate Report at 67.

[12] HSWA Sec. 224(a), adding RCRA §3005(a)(6), reflected at 40 C.F.R. §262.41.

[13] Senate Report at 67.

[14] HSWA Sec. 224(c), adding RCRA §8002(r).

[15] Senate Report at 68.

[16] 130 Cong. Rec. S9147 (daily ed., July 25, 1984).

Chapter 15

Solid Waste and Used Oil Recycling

Summary

Final Guidelines to Maximize Use of Recovered Materials Required:

- By May 1985, for the paper category.
- By October 1, 1985, for three additional categories (specifying tires as one).

Affirmative Program—Required Within One Year of Guidelines—Must Contain:

- Recovered material preference program.
- An agency program to promote the preference program.
- A program to estimate and verify percentage of recovered material in agency contracts.
- A program to award contracts based on percentage of recovered materials or, in the alternative, to specify minimum content of recovered materials.
- Annual review of the effectiveness of the agency preference program.

Final Decision to List Used Oil as Hazardous Required by November 8, 1986:

- Generators not subject to normal RCRA regulation if waste being properly recycled.
- Regulations must, while protecting human health and the environment, take into account effects on small businesses.
- Recycler deemed to have a permit if complies with Section 3004 standards.

Use of Recovered Materials by Federal Agencies

One of the principal objectives of the 1976 Act was resource recovery, that is the recycling of discarded wastes into fuel or usable products. According to Senator Jennings Randolph (D-West Virginia), then-Chairman of the Committee on Environment and Public Works, "America's massive quantities of solid waste constitute a virtually untapped supply of materials."[1]

To promote the worthy and popular objective of recycling some of this solid waste, the authors of the 1976 Act decided to harness the immense purchasing power of the federal government. Under Section 6002, procurement officers in every federal agency were directed to revise regulations to remove any discriminatory product specifications that would preclude the purchase of recycled products. The specific mandate of Section 6002 was to "procure items composed of the highest percentage recovered materials practicable consistent with maintaining a satisfactory level of competition."

Section 6002 also required EPA to develop guidelines for agencies to follow in procurement of five product categories. All of the guidelines were to be published by September 30, 1982. However, EPA produced only one guideline, which dealt with procurement of cement containing fly ash.[2] The slow pace at EPA was delayed even further by the Office of Federal Procurement Policy, which the House Energy and Commerce Committee labeled as "almost totally unresponsive"[3] to the congressional directive on resource recovery.

Frustrated by the Executive Branch's inaction, the Energy and Commerce Committee adopted an amendment to H.R. 2867 offered by Congressman Ron Wyden (D-Oregon) which revised Section 6002 and extended its deadlines. The Wyden amendment, after minor changes made on the House floor, was passed by the House on November 3, 1983. On July 25, 1984, the Senate voted to include a provision that was virtually identical to the Wyden amendment. The House language was adopted by the Conference Committee without discussion. Under the amended Section 6002, EPA was required to prepare a final guideline for paper within 180 days of the Amendments' enactment and for three additional product categories (including used tires) by October 1, 1985.* Also, within one year after EPA issues a guideline for products that can be made from recycled materials, each procuring agency must develop an "affirmative procurement program" which will assure that items composed of recovered materials will be purchased to the maximum extent practicable and which is consistent with applicable provisions of federal procurement law."[4] Each procurement program must contain a minimum of four basic elements:

1) a recovered materials preference program;

2) an agency program to promote the preference program;

3) a program for estimating the total percentage of recovered material used in the performance of a contract; certification of the recovered material

*EPA failed to meet these deadlines.

content actually used; and verification procedures for the estimates and certifications; and

4) annual review and monitoring of the effectiveness of an agency's affirmative procurement program.

The new provision underscored Congress' view that merely "encouraging" the purchase of recycled products was not enough.

Although there are numerous products that can be made from recycled or recovered materials, Congress was particularly concerned with promoting the use of retreaded tires, recycled paper, and the rubber from scrap tires for road surfacing. As waste, both used tires and discarded paper contribute heavily to the overcrowding of municipal solid waste facilities.

On April 9, 1985, EPA proposed a "Guideline for Federal Procurement of Paper and Paper Products Containing Recovered Materials."[5] Under the guideline, for purchases over $10,000, procuring agencies would solicit bids from all types of vendors including, specifically, those selling paper and paper products that contain "postconsumer recovered materials."* Vendors would be required to estimate the percentage of postconsumer recovered materials in their products. Preference would be granted to the lowest bid, but in the case of identical low bids, preference would be granted to the item containing the highest percentage of postconsumer recovered materials. In addition, if a paper product containing postconsumer recovered materials is consistently offered at a competitive price, EPA has recommended that the procuring agency adopt a minimum recovered materials standard for that item.[6]

Perhaps the most difficult aspect of the procurement program is the problem of verification. EPA pointed out that scientific verification of fiber content is impossible because virgin and recovered fibers possess the same chemical composition. However, the idea that procuring agencies conduct on-site inspections at the mill during the manufacturing process was rejected as "onerous, expensive and generally unnecessary."[7] And EPA regarded the suggestion of self-certification by the mill as "burdensome."[8]

On February 20, 1986, EPA proposed a "Guidelines for Federal Procurement of Asphalt Materials Containing Ground Tire Rubber for Construction and Rehabilitation of Paved Surfaces." The Agency noted that 240 million car and truck tires are discarded each year and that a wide range of recycling alternatives is available. According to EPA, scrap tires can be ground and used as binders and fillers for asphalt concrete, sealants, water proof pavement underlayers, bridge deck membranes, and seal coats. EPA advised procuring agencies that 'they can no longer exclude recycled materials" from these products unless the recycled materials jeopardized the "intended end use" of such products. 51 Fed. Reg. 6205.

*This term as defined by RCRA Section 6002(h) includes "paper, paperboard and fibrous wastes that have passed through their end usage as a consumer item" and "that enter and are collected from municipal solid waste."

Management and Recycling of Used Oil

In considering regulation of used oil, Congress in the 1984 Amendments addressed a problem it thought it had solved in 1980. In the Used Oil Recycling Act of 1980,[9] Congress directed EPA to develop a regulatory program for recycling used oil in accordance with certain congressional findings. Specifically, Congress found that:

> "1) used oil is a valuable source of increasingly scarce energy and materials;
>
> 2) technology exists to re-refine, reprocess, reclaim, and otherwise recycle used oil;
>
> 3) used oil constitutes a threat to public health and the environment when reused or disposed of improperly; and that, therefore, it is in the national interest to recycle used oil in a manner which does not constitute a threat to public health and the environment and which conserves energy and materials."[10]

The legislative history of the findings provision underscored Congress' view "that the recycling of used oil will result in the conservation of a valuable energy resource as well as diminish the likelihood of posing a threat to the public health and the environment if disposed of improperly."[11]

In January 1981, EPA said in a report to Congress that it intended to list both used and unused waste oil as hazardous wastes. However, during the first term of the Reagan Administration, the agency took no further regulatory action. In EPA's view, the blame for its failure to regulate waste oil belonged to the Congress. According to EPA, Congress omitted the directive to "protect human health and the environment" when it required EPA to ensure that its regulations did not discourage the recovery or recycling of used oil. Even though the 1980 Act clearly had incorporated the protection of human health and the environment standard, EPA's Office of General Counsel concluded that, because the seemingly indispensable phrase was missing from the appropriate place in the paragraph, the agency could not proceed with regulations. Consequently, the 1984 Amendments expanded the 1980 Act by adding 10 words (indicated by italics below) at the end of the final sentence so that the provision concerning EPA's regulatory duties now reads, in pertinent part, as follows: "[T]he Administrator shall promulgate regulations establishing such performance standards and other requirements as may be necessary to protect the public health and the environment from hazards associated with recycled oil. . . . The Administrator shall insure that such regulations do not discourage the recovery or recycling of used oil, *consistent with the protection of human health and the environment*."[12]

In addition to this slight modification of the Used Oil Recycling Act, Congress in 1984 enacted a parallel and, arguably duplicative, provision requiring the EPA Administrator to decide whether to propose listing or identifying used automobile and truck crankcase oil as a hazardous waste by

November 8, 1985, and to make a final determination by November 8, 1986.[13] In addition, Congress mandated that the regulations to be issued governing the recycling of used oils must be protective of human health and the environment and take into account the effect such regulations would have on small quantity generators and generators which are small businesses. To encourage recycling, Congress provided that even if the Administrator listed or identified used oil as a hazardous waste, generators of used oil would be exempt from most of RCRA's normal regulatory standards *provided the oil was properly recycled.*[14] Congress specified that the exemption is available only to a generator who either enters into an agreement with a transporter or an agent of the recycler for delivery of the used oil to a recycling facility with a valid permit or who recycles the used oil himself at a facility with a valid permit.[15] This exemption would not be available if the generator mixed the used oil with other types of hazardous waste, or failed to maintain basic records concerning recycling contracts and activities.[16] Congress also directed EPA to require transporters to deliver used oil only to facilities operating with a permit.

Under the 1984 Amendments, a used oil recycler would be automatically deemed to have a permit for treatment and recycling activities provided it complied with standards promulgated by the Administrator under Section 3004. However, Congress provided the Administrator with the authority to issue "individual permits" under Section 3005(c) "if he determines that an individual permit is necessary to protect human health and the environment."[17]

The used oil provisions of the 1984 RCRA Amendments represent a rather precarious regulatory balancing act, seeking to encourage recycling while protecting public health. As the conferees pointed out in the Conference Report:

> There is abundant evidence of harm due to improper recycling of used oil, such as incidents at Times Beach, Missouri, involving dioxin-contaminated used oil sprayed onto roads; incidents of used oil contaminated with other hazardous wastes being burned in large quantities for heating purposes; and the many used oil recycling sites on the list of Superfund National Priority Sites. At the same time, however, used oil, when properly recycled, can be a valuable resource . . . [18]

EPA's Proposal on Management and Recycling of Used Oil

On November 29, 1985, EPA proposed a comprehensive package of regulations in which the agency stated its intent to list many forms of used oil as a hazardous waste and to establish regulatory standards governing the recycling of used oil.[19] The proposal to list used oil as a hazardous waste was based on EPA's determination that used oil generally contains certain toxic contaminants such as lead,* cadmium, arsenic, chromium, toluene, naphthalene and chlorinated aliphatic hydrocarbons and therefore meets the

*The lead content of crankcase oil is directly related to the amount of lead in gasoline. Consequently, EPA's regulatory efforts under the Clean Air Act to reduce the amount of lead in gasoline is expected to have an equivalent effect on the presence of lead in used crankcase oil.

criteria under Section 3001.[20] The proposed rule would cover used oil as petroleum-derived and synthetic lubricants, hydraulic fluids, metal working fluids, insulating fluids, as well as sludges from the storage or processing of used oil. Certain materials would not be listed as used oil, including re-refined oil, oil rags, wastes from petroleum refining operations and waste waters contaminated with small amounts of used oil.[21]

In addition to listing certain classes of used oil as a hazardous waste, EPA's proposed rules would establish management standards governing generators, transporters, recyclers, and ultimate users of used oil. With respect to generators, EPA's proposal would create a distinction between those producing or storing less than 1,000 kilograms (about 300 gallons) per month and those who produce or store more. The "small quantity generators" of used oil would be exempt from most regulatory requirements, including manifesting and recordkeeping, provided the used oil is sent to a legitimate off-site recycler (or the generator recycles it himself) and no more than 1,000 kilograms is accumulated at one time. In addition, the exemption is only available if the used oil is not mixed with a hazardous waste.

Some regulatory restrictions *would* apply to small quantity generators: for example, the used oil could not be used as a dust suppressant on roads. In addition, if the used oil is stored in underground tanks, the tanks would have to meet the basic design standards established in Title I of RCRA (e.g., cathodic protection).[22]

Substantially more stringent regulatory requirements, similar to the rules applicable to other hazardous wastes, would be imposed on the large quantity used oil generators. However, these generators would not have to fill out a manifest form for each shipment so long as the used oil is sent to a permitted used oil recycling facility. The generators would, however, be required to maintain certain basic information about each shipment, including the EPA Identification Number of the recycling facility, for three years after each shipment.

Transporters would have to comply with similar recordkeeping requirements and would have to meet all applicable requirements concerning reporting accidents during shipping. In addition, they would not be allowed to store used oil at a transfer facility for more than 35 days, and they could not take more than 35 days to move the oil from a generator to a permitted recycling facility. Also, as required by the 1984 Amendments, the transporter must deliver the oil to the facility designated by the generator.[23]

The recycling facilities, which would include re-refiners, blenders, reprocessors and burners, would be deemed to have a RCRA permit if they complied with the standards set forth in Section 3004. However, such facilities would not be deemed to have a permit if the used oil is stored or treated in

*Under the proposed rule, used oil containing in excess of 1,000 parts per million of halogens will be presumed to have been mixed with hazardous waste. Unless such a presumption is successfully rebutted, used oil containing 1,000 ppm of halogens must be managed as a hazardous waste, not as recycled oil. 50 Fed. Reg. 49236 (November 29, 1985).

a surface impoundment or "used in a manner constituting disposal." Moreover, facilities that handled hazardous wastes in addition to used oil would have to be permitted on an individual basis.

Under EPA's proposed rule, recycling facilities handling non-specification used oil would generally be subject to the Section 3004 standards that are applicable to hazardous waste treatment, storage and disposal facilities. These standards will include some fairly stringent requirements—such as secondary containment—for storage and treatment in tanks. In addition, EPA has proposed a number of other requirements that are specifically related to oil recycling operations. For example, all oil recyclers would have to test used oil for halogens, flashpoint and for the other criteria for "specification fuel." [See chart in Chapter 11, page 143.]

An issue that EPA (and perhaps Congress) will need to address involves the impact that listing used oil as a hazardous waste will have on its marketability. During the debate on S. 757, Senator Chafee stated: "While there have been some isolated problems, in general the waste oil industry has demonstrated how a potentially hazardous substance can be safely collected and profitably recycled into new, useful products. We do not want to upset an industry that already is doing voluntarily what we now are requiring others to do."[24] The problem that Congress did not consider in enacting the used oil management provision is that once used oil is designated a hazardous waste under RCRA, it also becomes a hazardous substance under Superfund. This status undoubtedly will have the effect of lowering the value of the oil, because the dual listing means that those handling used oil (generators, transporters and recyclers) would be subject to Superfund's joint, several and strict liability provisions. Obviously, the prospect of Superfund liability will be a factor that cannot be ignored,* particularly because the delicately balanced recycling system requires each participant in the system to fulfill its particular task.

Simply stated, the dilemma EPA faces is this: When the market incentives to manage used oil as a product decrease, the problems of managing it as a waste will increase. As long as there is a recycling system in place, there is a way of handling the approximately one billion gallons of waste oil produced each year in America. On the other hand, the more burdensome the requirements, the less likely it is that this massive quantity of material will be channelled into proper recycling uses. For example, if the price differences are small, burners may simply turn to virgin oil as fuel and dispense with the liability risk of burning a "hazardous waste." Yet the problem of *where* the used oil would go would remain. The likely candidates are landfills, sewers, and backyards.

On May 19, 1986, during a hearing on the used oil issue held by a House Small Business Subcommittee, EPA Assistant Administrator J. Winston Porter admitted that the regulatory impact analysis used by the Agency to support

*On December 6, 1985, the House of Representatives adopted an amendment to the Superfund reauthorization bill which would, in essence, preclude Superfund liability for used oil that was being recycled in accordance with EPA's regulations. See 131 Cong. Rec. H11620 (daily ed., Dec. 10, 1985).

its November 29 proposal was no longer valid. He also indicated that the overwhelmingly negative response to the proposed regulations had caused the Office of Solid Waste to reassess its options, including one in which recycled used oil would be regulated under the Used Oil Recycling Act but not listed as a hazardous waste. Porter did not mention that option has been available since 1980.

Epilogue

Just prior to its November 8, 1986 deadline, EPA determined that recycled used oil should not be listed as a hazardous waste. The Agency's decision was based on the realization that listing would discourage recycling activities which in turn "could cause an increased quantity* of used oil to be disposed of in uncontrolled ways, causing harm to the environment." According to EPA, such increased disposal would "result from decreased use of used oil as fuel by industrial burners and decreased acceptance of do-it-yourselfer oil by service stations (and similar establishments) both attributable to costs and stigma associated with listing." Thus, following years of delay and debate, EPA had become convinced that recycled oil "occupies a unique position in the Subtitle C structure and is to be accorded distinct regulatory treatment."

The Agency's decision not to list was aided to some extent by amendments to RCRA contained in the Superfund Amendments and Reauthorization Act (Public Law 99-499) which was signed into law by President Reagan on October 17, 1986. The amendments made two important changes to the used oil regulatory framework. First, section 3006 was amended to provide EPA with the ability to authorize states to implement recycled oil regulatory programs whether or not used oil was a listed or identified hazardous waste. Second, improper handling or disposal of used oil was made subject to the criminal provisions set forth in section 3008(d).

The other provisions relating to used oil in the 1986 Superfund Amendments create a limited exemption from Superfund liability for "service station dealers"** who meet certain criteria. To qualify for the exemption, which applies only to off-site liability, a service station dealer must comply with used oil management standards under section 3014 and offer a collection facility for do-it-yourselfer used oil. In addition, the used oil may not be mixed with any hazardous substance.

*A November 1986 report, by Temple, Barker and Sloane, Inc., cited by EPA, estimated that listing recycled oil as a hazardous waste could cause an additional 61 to 128 million gallons of used oil to be disposed of improperly.

**Service station dealer is defined in section 114 of the 1986 Superfund Amendments as an owner or operator of "a motor vehicle service station, filling station, garage or similar retail establishment . . . where a significant percentage of the gross revenue is derived from the fueling, repairing or servicing of motor vehicles." Service station facilities owned by the government also qualify.

Chapter 15 Footnotes

[1] 122 Cong. Rec. S21401 (daily ed., June 30, 1976).
[2] 40 C.F.R. Part 249; 48 Fed. Reg. 4230 (Jan. 28, 1983).
[3] House Report No.198, 98th Cong., 1st Sess., Pt.1 at 70 (May 17, 1983).
[4] HSWA Sec. 501(a), adding RCRA §6002(i).
[5] 50 Fed. Reg. 14076.
[6] 50 Fed. Reg. 14079.
[7] 50 Fed. Reg. 14080.
[8] *Id.*
[9] Public Law 96-463.
[10] Section 1 of Used Oil Recycling Act of 1980, 42 U.S.C. §6901(a).
[11] House Rep. No. 1415, 96th Cong., 2nd Sess. 10 (Sept. 26, 1980).
[12] HSWA Sec. 242, amending RCRA §3014(a).
[13] HSWA Sec. 241(a), adding RCRA §3014(b).
[14] HSWA Sec. 241(a), adding RCRA §3014(c)(1).
[15] HSWA Sec. 241(a), adding RCRA §3014(c)(2)(B).
[16] HSWA Sec. 241(a), adding RCRA §3014(c)(2)(B)(ii) and (iii).
[17] HSWA Sec. 241(a), adding RCRA §3014(d).
[18] Conference Report at 113.
[19] 50 Fed. Reg. 49202 *et seq.*.
[20] See 40 C.F.R. §261.11(a)(3).
[21] 50 Fed. Reg. 49261 (Nov. 29, 1985).
[22] EPA's proposed regulations for underground storage tanks containing hazardous wastes are set forth in 50 Fed. Reg. 28444 (June 26, 1985).
[23] 50 Fed. Reg. 49234 (Nov. 29, 1985).
[24] 130 Cong. Rec. S9193 (daily ed., July 25, 1984).

Chapter 16

The Enforcement Provisions

Summary

Imminent and Substantial Endangerment Provision:

- Clarified to assure application to present releases of previously disposed hazardous wastes and the generators of those wastes.
- Citizens authorized to file imminent and substantial endangerment suits, except where government is diligently pursuing an enforcement action or where a site is being cleaned up.

Corrective Action:

- EPA authorized to compel corrective action at interim facilities through administrative order and lawsuits, rather than only as part of permit.
- EPA may require corrective action beyond boundary lines where necessary to protect health.

Criminal Provisions:

- Lower burden of proof and increased penalties for the "knowing endangerment" offense.
- Both generator and transporter liable for midnight dumping.
- Illegal export of hazardous waste made a criminal offense.
- Omission of material information to EPA or a state made a criminal offense.
- Violation of interim status regulations made a criminal offense.
- Penalties increased.

Export of Hazardous Waste Requires:

- A bilateral agreement between the U.S. and the receiving country, or
- Various notifications and consents.

Inspections and Inventories:

- By May 8, 1985, EPA must submit a report on the merits of using private inspectors to supplement government inspectors.
- By November 8, 1985, EPA must commence a program of inspecting all hazardous waste facilities at least every two years.
- By November 8, 1985, EPA must undertake an annual inspection of each federal hazardous waste facility.
- Every two years, each federal agency must submit to EPA an inventory of its hazardous waste facilities.

EPA Ombudsman:

- Must be established to act on complaints and information requests concerning RCRA.

Congress' preoccupation with closing RCRA loopholes was nowhere more evident than in the detailed attention it gave to strengthening and improving the Act's enforcement provisions. Congress was well aware that the more stringent substantive provisions of the 1984 RCRA Amendments would make enforcement even more critical. According to the Senate Report, "[A]s the implementation of other provisions of this bill restrict land disposal of hazardous wastes and require safer methods of handling and treatment, there will be a significantly greater incentive to dispose of toxic waste illegally."[1] Consequently, the enforcement provisions of "the new RCRA" not only dramatically expand the scope of liability of the regulated community, but also increase the government's ability to successfully prosecute violators and obtain more severe punishment. If there are any enforcement loopholes left in RCRA, it is because they were not brought to Congress' attention.

Revisions to the 'Imminent and Substantial Endangerment' Provision

Of all the changes made to the civil enforcement provisions, perhaps the most important is the "clarification" of the scope of Section 7003, the "imminent and substantial endangerment" provision. When Congress originally passed RCRA in 1976, it relied primarily on regulatory controls. However, Congress also provided EPA with the ability to obtain injunctive relief against anyone contributing to an imminent and substantial endangerment created by the handling, storage, treatment, transportation, or disposal of any solid or hazardous waste. Since 1979, the Department of Justice, on behalf of EPA, has filed approximately 90 Section 7003 actions. Yet, despite its extensive and usually successful use, Section 7003 had received some rough treatment at the hands of four federal district courts.

The first adverse decision, from the government's point of view, resulted from litigation over the Wade disposal site in Chester, Pennsylvania. After a fire at the site in February 1978, EPA conducted an investigation which revealed that the site contained several thousand ruptured, charred and corroding drums, along with several tank trucks and trailers from which toxic and explosive wastes were leaking. EPA's investigation also revealed that some of the wastes delivered to the site had been dumped directly onto the ground and into drains which discharged into the Delaware River.

In 1979, the Justice Department filed on behalf of EPA a Section 7003 action against the owner and operator of the Wade site and the company which transported the wastes to the site. In 1981, the United States amended its complaint to add claims against six companies that generated wastes disposed of at the site. On September 7, 1982, the district court dismissed the complaint against the generators.[2]

In its ruling, the court assumed that the site did in fact present an imminent and substantial endangerment and that the generators were the source of at least some of the wastes at the site. Nevertheless, the court concluded that, because Section 7003 "is written in the present tense" and because its operative language authorizes suits "to stop" and "to immediately restrain" certain activities, the section could apply only to ongoing activities.[3] Although RCRA defines "disposal" to include "leaking"[4], the court "inclined to the view that current leaking of previously dumped waste does not constitute 'disposal'."[5]

The court also rejected the government's argument that Section 7003 grants it authority to enjoin a past off-site generator as a "person contributing to" the activity which caused the endangerment. The court stated that there was "no logical limit, given the breadth of the statutory language, to the number and type of persons who might be construed as 'contributing to' the disposal of hazardous waste." Therefore, without a definition of the phrase "any person contributing to," or other congressional guidance, the court concluded that it would be unreasonable to read the term so as to include such generators.[6]

A similar ruling was issued in *United States v. Northeastern Pharmaceutical and Chemical Company, Inc.*[7] In that case, the court acknowledged that the disposal of wastes containing high concentrations of hexachlorophene, toluene, and dioxin at the Denney farm site in Missouri had resulted in an imminent and substantial endangerment to human health and the environment. However, the court, noting that the government had not alleged negligence on the part of the defendants, ruled that Section 7003 "does not apply to past non-negligent off-site generators or transporters." The court based its holding in part on a very curious reading of legislative history:

> Although it could be argued that Congress was unaware of the problems arising from inactive or abandoned sites, Congress expressly recited such instances of the damage caused by hazardous waste disposal practices in the legislative history. [H.R. Rep. No. 1491, *supra*, at 17-24, *reprinted in* (1976) U.S. Code Cong. & Ad. News at 6254-6261.] It would be more properly stated that Congress was unaware of the magnitude and expense of inactive or abandoned sites, as well as the lack of current means or financially responsible parties to clean up those sites. This Court concludes that Congress, knowledgeable of the existence of hazardous waste problems, chose to principally direct RCRA's provisions toward the regulation of the source and not the results of hazardous waste disposal.[8]

With respect to the issue of negligence, the court focused on an ambiguous paragraph in a 1979 report by the Oversight and Investigations Subcommittee of the House Interstate and Foreign Commerce Committee (later renamed the Energy and Commerce Committee). That paragraph, unfortunately, demonstrated considerable confusion on the part of the Subcommittee regarding the doctrine of strict liability:

> [S]ection 7003 should not be construed solely with respect to the common law. Some terms and concepts, such as person "contributing to" disposal resulting in a substantial endangerment,

> are meant to be more liberal than their common law counterparts. For example, a company that generated hazardous waste might be someone "contributing to" an endangerment under Section 7003 even where someone else deposited the waste in an improper disposal site (similar to strict liability under common law), where the generator *had knowledge* of the illicit disposal or failed to exercise due care in selecting or instructing the entity actually conducting the disposal.[9] [Emphasis added.]

Apparently, the court, based on its reading of this excerpt, concluded that while strict liability might be the appropriate standard for on-site generators, a negligence standard is applicable to off-site generators.

Although a number of cases offered an expansive reading of the scope of Section 7003,[10] the 98th Congress, as it commenced the RCRA reauthorization process, was primarily concerned with the *Wade* and *Northeastern Pharmaceutical* rulings and similar holdings in *United States v. Waste Industries, Inc.,*[11] and *United States v. Midwest Solvent Recovery Inc.*[12]

In reporting H.R. 2867, the Energy and Commerce Committee adopted revisions to Section 7003 which it described as simply a "clarification" of existing law. The Committee left no doubt, however, that it intended to legislatively overrule the interpretations handed down by the district courts in *Wade* and *Waste Industries*:

> These amendments are intended to clarify the breadth of Section 7003 as to the persons, conditions and acts it covers. . . . [A]nyone who has contributed to the creation, existence or maintenance of an imminent and substantial endangerment is subject to the equitable power of Section 7003, without regard to fault or negligence. Such persons include, but are not limited to, past and present generators (both off-site and on-site) . . . past and present owners and operators of waste treatment, storage or disposal facilities and past and present transporters . . . Thus, for example, non-negligent generators whose wastes are no longer being deposited or dumped at a particular site may be ordered to abate the hazard to health or the environment posed by the leaking of wastes they once deposited or caused to be deposited on the site.[13]

The report on S.757 by the Senate Environment and Public Works Committee contained virtually identical language to describe the same changes to Section 7003.[14]

The Senate Report also dealt with two other litigation issues that arose under Section 7003. First, the Report emphasized that "an endangerment means risk of harm, not necessarily actual harm" and that "recognizing the risk may be assessed from suspected, but not completely substantiated, relationships between facts, from theoretical projections, from imperfect data, or from probative preliminary data not yet certified as fact." [Senate Report at 59 quoting *United States v. Vertac Chemical Corp.*, 489 F. Supp. 870, 885 (E.D. Ark. 1980).] Second, the Report asserted that United States is not required to pursue all its alternatives to injunctive relief (e.g., under Superfund) before filing a Section 7003 action. According to the Senate Report, Section 7003 is a free-standing provision which may be used as an "alternative and supplement to other remedies."

The final version of the bill retained the changes approved by both the House and the Senate, and the Conference Report repeated much of the discussion of Section 7003 which appeared in the House and Senate Committee Reports. Congress thereby avoided the triple problem of silence, confusion, and ambiguity which had plagued the legislative history of Section 7003 in its original form. Congress wanted to remove any doubt that this provision (and its citizen suit Section 7002 counterpart) was meant to be a powerful weapon in the effort to compel cleanup of serious hazardous waste problems.

Citizen Suits

In addition to expanding the scope of Section 7003, Congress took steps to provide citizens with authority to bring "imminent and substantial endangerment" actions. Under the 1976 version of the citizen suit provision, citizens could sue to correct specific violations of RCRA's regulations. Under the new provision, citizens are given much broader rights to force responsible parties to clean up a hazardous or solid waste site that presents a significant danger to human health or the environment. However, in order to prevent these suits from interfering with ongoing actions being taken to remedy the problem, Congress prohibited a citizen from filing an imminent and substantial endangerment lawsuit whenever the government or responsible party was engaged in a clean-up action under Superfund, *or* when the government was "diligently prosecuting" an enforcement action against the responsible parties.

According to the House Report, the "expansion of the citizens suit provision will complement, rather than conflict with, the Administrator's efforts . . . , particularly where the Government is unable to take action because of inadequate resources."[15] The Energy and Commerce Committee also addressed another concern, namely that citizens would use the doctrine of "pendent jurisdiction" to bring tort claims under state law along with the federal Section 7002 action. Although the Committee declined to prohibit that practice, it assumed courts would "exercise their discretion concerning pendent jurisdiction in a way that will not frustrate or delay the primary goal of this provision, namely the prompt abatement of imminent and substantial endangerments."[16] Section 7002 was also revised to legislatively overrule a 1983 federal district court decision[17] which held that the prohibition on open dumping of solid waste could not be enforced through a citizens suit provision, unless the state had obtained EPA's approval of its Subtitle D solid waste program.[18]

Direct Action Against Guarantors

Ever since the enactment of Superfund in 1980, insurance companies had been more than a little concerned about a provision in Section 108(d) of CERCLA which could force an insurer who failed to act in "good faith" to pay the full amount of the environmental damage regardless of the limits on coverage specified in the policy. Although the Superfund statute does not define the meaning of either "good faith" or "bad faith," these terms had acquired meaning in case law in most states. Nevertheless, the insurance companies were worried that courts in Superfund cases would search for a "deep pocket" and therefore would construe these terms so as to find bad faith on the part of the insurer whenever possible and thereby "break" the limits of the policy.

To cure this uncertainty, insurance industry representatives came to the Senate Environment and Public Works Committee staff requesting a package of amendments including one which specified that an insurer's policy limits could not be exceeded. The proposal made clear that any other statutory, contractual, or common law liability of a guarantor (insurer) would not be affected. The package of amendments applied to both RCRA and Superfund and thus would have established uniform financial responsibility requirements in both statutes. As an incentive to Congress to adopt the entire package, the industry included a provision to allow the government to sue the guarantor directly to recover the costs of cleanup (up to the policy limits) in cases where the owner/operator has gone bankrupt, fled the country or is otherwise unavailable. The Senate adopted the package of amendments on July 25, 1984; but, ironically, the conferees adopted the "direct action" provision for RCRA only and deleted all Superfund amendments, leaving the insurance industry without the key change in the Superfund law that motivated it to seek the amendments in the first place.

As enacted, the direct action amendment, Section 205,[18] makes several changes to the financial responsibility provisions of RCRA. First, it specifies that a guarantor's policy limits will be binding. This provision would not, however, limit any other statutory, contractual, or common law liability of a guarantor, including an insurer's traditional common law obligation to act in "good faith" toward its insured. Second, Section 205 provides claimants with the right of "direct action" against guarantors. This means that where, for example, the facility is in bankruptcy or is not subject to state or federal court jurisdiction, the insurance company can be sued and the cost of the corrective action (or damages resulting from personal injury) can be collected. Third, the guarantor is allowed to invoke the terms and conditions set forth in the policy as defenses in a direct action suit. In authorizing a guarantor to invoke these rights and defenses (thus in some circumstances limiting a guarantor's liability), it was Congress' hope that the amendment would foster the development of a competitive marketplace for insurance, and thereby encourage greater availability of reasonably priced insurance. Unfortunately, this result is unlikely to be forthcoming in the foreseeable future.

Finally, the amendment authorized the Administrator of EPA to determine appropriate rules and defenses with respect to this type of action. According to the Conference Report, EPA would have the discretion to "specify policy or other contractual terms, conditions, or defenses that are unacceptable or are unnecessary to protect human health and the environment."[20]

Corrective Action at Interim Status Facilities

Another major expansion of EPA's enforcement authority conferred by the 1984 Amendments is the ability to require the owners and operators of "interim status" facilities to take "corrective action." Under EPA's old regulations, the agency believed that it could not compel cleanup of contaminated groundwater or the repair of leaks, except in the facility's final permit which could take years to complete.[21] Senator Daniel Patrick Moynihan (D-New York) argued that proceeding through the permit process was too slow an approach to the serious problem of leaking interim status facilities.[22] Echoing Moynihan's views, the Conference Report pointed out that although "the issuance of a land disposal permit can be very time-consuming," the reasons do not necessarily involve problems associated with groundwater contamination."[23] According to the conferees, "EPA should have the power to deal directly with an ongoing environmental problem without awaiting the issuance of a final permit."

Consequently Section 233 provides the agency with authority to compel corrective actions at interim facilities both through administrative orders and civil suits.[24] Noncompliance with EPA's administrative orders for corrective action are made subject to a civil penalty of $25,000 per day.[25]

Corrective Actions Beyond Facility's Property Line

The 1984 Amendments also overturn EPA's policy of limiting corrective action to within the property line of a leaking facility. As the Conference Report noted somewhat sarcastically, "[S]ince most forms of pollution, particularly groundwater contamination, do not observe territorial or property boundaries, such a restriction has no basis in logic."[26] Thus, under Section 207, EPA is directed, "as promptly as practical," to amend its regulations to require corrective action beyond the facility boundary line when such action is necessary to protect human health and the environment. However, corrective action would not be required where the owner or operator shows that, despite its best efforts, it is unable to obtain permission to undertake such action.[27] The regulations are to take effect immediately upon promulgation, and apply to all permitted facilities, interim status landfills, impoundments and waste piles that received waste after July 26, 1982. That is the date that EPA regulations, as of January 26, 1983, instituted a corrective action requirement for those facilities still operating. After those regulations were issued, some facilities had stopped operating in order to avoid the corrective action requirements that were to become effective in six months. The Amendments make the requirement applicable to these facilities.[28] Until EPA issues its new corrective action regulations, it must issue corrective action orders on a case-by-case basis as necessary to protect human health and the environment.

Criminal Provisions

On the criminal side of the enforcement equation, Congress plugged a number of loopholes and added other provisions designed to strengthen the government's prosecutorial efforts. Perhaps the most important change concerned the "knowing endangerment" provision, which makes it a felony to "knowingly place another person in danger of death or serious bodily injury" by improper handling or disposal of hazardous waste.

When Congress enacted this provision in 1980, it overloaded the government's burden of proof by requiring prosecutors to demonstrate that the defendant showed either an "unjustified and inexcusable disregard" for human life or an "extreme indifference" to human life. This "state of mind" requirement was in addition to the burden of proving the knowing violation of a basic requirement such as disposing hazardous wastes at a facility without a permit. Principally because of the complexity of the 1980 version of this provision and the extraordinary burden of proof imposed, only one indictment was handed down.* Clearly a criminal provision that is virtually impossible to prosecute does not provide much deterrence. Consequently, Congress simplified the provision by dropping the requirements concerning "disregard" or "indifference" to human life.[29] At the same time, the maximum prison sentence was tripled—from 5 to 15 years. The maximum fines remain the same: $250,000 for an individual, and $1 million for a company.

Other key changes to the criminal provisions were enacted to make sure that anyone responsible for midnight dumping activities could be prosecuted. Section 3008 was revised to allow the prosecution of a generator who knowingly causes its waste to be transported to a facility without a permit.[30] That would include, for example, a ditch by the side of the road. Previously, only the transporter could be found guilty.[31]

Other changes were made to reduce the chances that defendants would be acquitted on the basis of technicalities. It is now a criminal offense to knowingly omit material information that is required to be submitted to EPA or a state.[32] "Material omissions" are those which would tend to influence agency action. Previously, it was a crime only to falsify such information.

The 1984 Amendments also make it a crime to knowingly violate a material interim status regulation.[33] Prior to November 1984, it had been unclear

*More than a year after the enactment of the 1984 Amendments, a federal grand jury indicted Arthur J. Greer, the owner of four companies which handled and disposed of hazardous wastes in Florida, for the knowing endangerment of several of his employees. The indictment was based on events that occurred prior to the changes to the criminal provisions.

whether deliberate violations of the interim status rules could be prosecuted. Additionally, the 1984 Amendments increase the fines and sentences for criminal violations, making all offenses felonies, rather than misdemeanors which federal prosecutors frequently ignore. Each day of violation can be punished by up to a $50,000 fine and a prison sentence of up to two years for violations involving recordkeeping, false submissions, unmanifested transportation, and unauthorized exports, and five years for the other violations, such as storage or disposal without a permit.[34]

Also in the 1984 RCRA Amendments, Congress gave special emphasis to EPA's criminal investigators: first, by providing in a line-item authorization of $11 million for their work over the next four years and, second, by directing the Attorney General to provide them with "law enforcement" authority such as the ability to issue subpoenas, carry firearms, and make arrests.[35] Without such authority, EPA's investigators could not arrest a midnight dumper, even one they caught red-handed. (The House legislative battle over this "law enforcement" authority is discussed in detail in Chapter 22.)

Ever vigilant in its endeavor to close RCRA loopholes, Congress even pounced on a loophole that wasn't really there. The case of the non-existing loophole originated in Mount Laurel, New Jersey, at a plant owned by the Johnson & Towers Company, where trucks were overhauled and repaired.

On June 8, 9, and 10, 1981, EPA criminal investigators conducted a surveillance operation and observed some of the waste disposal practices at the Mount Laurel plant. Substantial quantities of degreasers containing chemicals such as methylene chloride and trichlorethylene, both of which EPA has listed as hazardous wastes, were dumped into a holding tank, and when the tank was full, the contents of the tank were pumped into a trench. The trench flowed from the plant property into Parker's Creek, a tributary of the Delaware River. After EPA's investigation, a federal grand jury indicted Johnson & Towers (which did not have a RCRA disposal permit) and two employees, Jack Hopkins, a foreman, and Peter Angel, the plant supervisor. The defendants were charged with three counts of violating RCRA and one count of violating the Clean Water Act. The RCRA portion of the indictment alleged that the defendants had knowingly stored and disposed of hazardous wastes without having obtained a permit. The Johnson and Towers Company pled guilty; the employees pled not guilty.

The defense offered by the lawyers for Hopkins and Angel was ingenious: Hopkins and Angel were not "persons" as defined by RCRA. Although the word "individual" was included within the statutory definition of "persons," Congress really meant the term "individual" to refer to "owner or operator," since it was the owner or operator of a waste disposal facility who had the obligation to obtain a RCRA permit. Because the individual defendants were not owners or operators and therefore did not have overall operating authority over the plant, they were beyond the reach of RCRA's criminal provisions. Amazingly, the federal judge hearing the case, Stanley Brotman, was persuaded by this argument. He dismissed the RCRA charges against Hopkins and Angel.

Although the Justice Department appealed the ruling, members of Congress were outraged. Congressman Florio quickly introduced a bill to "clarify" the language of Section 3008(d) to make clear that anyone violating RCRA's criminal provisions could be held liable. However, because Florio's bill was introduced after the House had passed H.R. 2867, any timely legislative overruling of the *Johnson & Towers* decision had to be done in the Senate.

Consequently, Florio asked New Jersey's senators to offer a Senate floor amendment to S. 757. On July 25, 1984, Senators Frank Lautenberg and Bill Bradley, both Democrats, persuaded their colleagues to remove any possible doubt as to the scope of Section 3008(d). Like the Johnson & Towers method of hazardous waste disposal, the Lautenberg-Bradley amendment was not elaborate. It simply removed the words "having obtained" before "a permit," so that the provision would read: "Any person who knowingly . . . disposes of any hazardous waste . . . without a permit . . .", is guilty of a criminal offense. Excising "having obtained" eliminated the contention that only those who were required to obtain permits (owners and operators) could be held responsible for RCRA violations.

Explaining the importance of the change, Senator Lautenberg said that the amendment "will close a dangerous loophole created by a recent federal district court decision that could let many midnight dumpers of hazardous waste off the hook."[36] Lautenberg went on to explain: "The language in the law is clear and unambiguous. It imposes a criminal penalty on any person who, lacking a permit, knowingly and illegally treats, stores, transports, or disposes of any hazardous material. It was not the intent of Congress that criminal liability extend only to the owners and operators of facilities, who are responsible for obtaining a RCRA permit. This, however, is the way the district court interpreted RCRA."

Senator John Chafee of Rhode Island, the floor manager of S. 757, expressed his bewilderment with the court's decision:

> I must say that I do not know how the Federal judge ever arrived at his conclusion . . . How this could have happened leaves me dumbfounded. Nonetheless, it did happen and it seems wise to correct it.[37]

The Lautenberg-Bradley amendment was adopted without further debate. Less than a month later, the United States Court of Appeals reversed the district court's dismissal of the indictments against Angel and Hopkins. The Appeals Court stated that "under the plain language of the statute the only explicit basis for exoneration is the existence of a permit covering the action. Nothing in the language suggests that we should infer another provision exonerating persons who knowingly treat, store or dispose of hazardous wastes but are not owners or operators."[38]

Export of Hazardous Waste

Before enactment of the 1984 Amendments, "midnight dumping" could be accomplished by exporting hazardous wastes abroad. There was no requirement that foreign countries receiving hazardous wastes be apprised of the nature of the shipment. Nor was there any obligation under federal law that the exporters obtain the consent of the receiving country.[39]

The "export" loophole was a potential problem that Congresswoman Barbara Mikulski (D-Maryland) believed should be addressed because restrictions on land disposal would make the option of exporting hazardous waste more attractive. As a result of the Mikulski amendment offered on the House floor in October 1983, RCRA now requires that, as of November 8, 1986, no person may export hazardous waste unless:

1) a notification has been filed with EPA;
2) the receiving country has agreed in writing to accept the waste;
3) a copy of the consent is attached to the manifest; and
4) the shipment conforms to the terms of the consent.[40]

On March 13, 1986, EPA issued proposed regulations to implement these requirements.[41] These export requirements, set forth in Section 245, are not applicable where there is a bilateral agreement, between the United States and the government of the receiving country, which establishes hazardous waste export procedures, and where shipments conform with the terms of the agreement.[42] The Senate Report states that the bilateral agreement should describe joint efforts to monitor and spot-check shipments of hazardous waste as well as the respective responsibilities for enforcement and prosecution of the terms of the agreement.[43]

As to the notification requirement for shipping hazardous wastes, EPA in its proposed regulations has determined that notification is required 60 days before the initial shipment is intended to be shipped off site. This notification may cover export activities extending over a 24-month or lesser period.[44] The Amendments also require that exporters submit annual reports of their hazardous waste exports.[45] EPA has implemented this provision in regulations requiring that, by March 1 of each year, a report and a summary of the types, quantities, frequency and ultimate destination of all hazardous waste exported during the previous year be submitted to EPA.[46] The Senate Report states that the public should be given full access to information produced under the notification provision.[47]

Finally, the Amendments provide criminal penalties of a fine of not more than $50,000 or imprisonment of up to two years, or both, for knowingly exporting hazardous waste without the consent of the receiving country, or in violation of an existing international agreement between the United States and the receiving country.[48]

Ombudsman

In response to the general concern that citizens were unable to get assistance from EPA, Congress established an Office of Ombudsman within EPA charged with acting on individual complaints and requests for information concerning the Resource Conservation and Recovery Act.[49] The House Energy and Commerce Committee Report described the problem in this way: "The American public, particularly those communities close to hazardous waste facilities, often have been frustrated in their attempts to obtain information concerning the health danger posed by improperly disposed hazardous waste. EPA has been hampered in its ability to communicate with the public by not having a single office whose essential purpose is to respond to citizen inquiries and complaints."[50]

During the conference, Congresswoman Barbara Mikulski was the provision's chief proponent and Senator Chafee its chief critic. Senator Chafee and other Senate conferees felt that EPA already had a responsibility to respond to the public's concerns, and they were reluctant to expand EPA's bureaucracy in this manner. As part of the final compromise on the issue, the conferees agreed to establish the Office of Ombudsman, but not on a permanent basis—it is due to expire on November 8, 1988. The Conference Report states that the individual appointed to be Ombudsman should be of sufficient status within the agency that citizens will be able to secure meaningful assistance as quickly as possible. The Conference Report also states that there should be staff resources for the Ombudsman Office at EPA headquarters and at each of the regional offices.[51] More than a year after the enactment of the Amendments, Robert Knox, a career EPA employee with considerable RCRA expertise, was appointed Director of the Office of Ombudsman.

Mandatory Inspection and Inventory

Since enactment of RCRA in 1976, EPA and the states have had clear authority to inspect hazardous waste facilities. However, because they were using this authority so infrequently[52] (with most facilities not being inspected at all), Congress decided in the 1984 Amendments to impose mandatory inspections.[53]

Section 231 requires that by November 8, 1985, EPA must commence a program of inspecting every treatment, storage, and disposal facility at least every two years.[54] The House Report pointed out that "[a] mandatory program of periodic inspections . . . should provide substantial incentives for voluntary compliance."[55] Congress was concerned about the quality as well as the frequency of inspections. In both the Senate and House reports, Congress directed EPA to address this concern in its regulations and to consider requiring that all inspections be conducted by an inspector certified by EPA as competent.[56]

Section 231 also requires that by May 8, 1985, EPA must submit to Congress a report on the merits of using private inspectors to supplement government inspections.[57] Although Congress considered private inspectors a possible solution to the problems posed by inadequate state and federal resources, it recognized potential problems with their use, and directed EPA to consider these in its report. Among the problems specifically mentioned were potential conflicts of interest and the need to keep the inspection data confidential.[58] Finally, both the Senate and House Reports directed EPA to consider establishing a schedule of fees to be charged the companies for the inspections.[59]

During the debate on this issue, Congresswoman Barbara Mikulski wanted to know "who was going to enforce the enforcers?"* In adopting her amendment requiring the inspection of state and federal facilities, Congress attempted to assure that the enforcers of the law were not themselves allowed to evade it.[60] In addition to requiring annual inspections of state and federal facilities, Congress required that EPA conduct the inspection of federal facilities. Although EPA's inspection duty cannot be assumed by states even when they have authorized programs,[61] nothing in the law precludes authorized states from conducting their own inspections. In addition, the Amendments require that the records of all inspections of all state and federal facilities be made available to the public.[62]

Finally, the Amendments require that every two years each federal agency must submit to EPA and to authorized states an inventory of each treatment, storage, or disposal facility that it owns or operates or previously owned or operated.[63] Agencies need not resubmit information already submitted under Section 103 of CERCLA or Sections 3005 and 3010 of RCRA. If a federal agency fails to conduct the required inventory, EPA is required to do so.[64] The Senate Environment and Public Works Committee, which was concerned that federal agencies and EPA would ignore their obligations to compile an inventory, pointed out that these nondiscretionary duties would be enforceable by citizens under Section 7002.[65]

*Mikulski's concern was prompted in large measure by the Monument Street landfill in Baltimore—a disposal site operated by the city and now a Superfund site. See, Hearings before the Subcommittee on Commerce, Transportation and Tourism, Committee on Energy and Commerce, No. 154, 97th. Cong., 2nd. Sess., 145 (May 15, 1982).

Chapter 16 Footnotes

[1] Senate Report at 46.
[2] *United States v. Wade*, 546 F. Supp. 785 (E.D. Pa. 1982).
[3] 546 F. Supp. at 789.
[4] RCRA §1004(3).
[5] 546 F. Supp. at 790.
[6] *Id.*
[7] 579 F. Supp. 823 (W.D. Mo. 1984).
[8] 579 F. Supp. at 835.
[9] House Report No. 172, 96th Cong., 2d Sess. 5, *reprinted in* 1980 U.S. Code Cong. & Ad News 5019, 5023 (emphasis added), cited at 579 F. Supp. at 236.
[10] E.g., *United States v. Price*, 688 F.2d 204 (3d Cir. 1982); *United States v. Solvents Recovery Service of New England*, 496 F. Supp. 1127 (D. Conn. 1980).
[11] 556 F. Supp. 1301 (E.D.N.C. 1982).
[12] 484 F. Supp. 138 (N.D. Ind. 1980).
[13] House Report at 56.
[14] Senate Report at 58.
[15] House Report at 53.
[16] *Id.*
[17] *City of Gallatin v. Cherokee County*, 563 F. Supp. 940 (E.D. Texas 1983).
[18] HSWA Sec. 401(a)(1), amending RCRA §7002(a)(1).
[19] Adding RCRA §3004(t).
[20] Conference Report at 91.
[21] Conference Report at 110, 111.
[22] 130 Cong. Rec. S9175 (daily ed., July 25, 1984).
[23] Conference Report at 111.
[24] HSWA Sec. 233(a), adding RCRA §3008(h)(1).
[25] HSWA Sec. 233(a), adding RCRA §3008(h)(2).
[26] Conference Report at 92.
[27] HSWA Sec. 207, adding RCRA §3004(v).
[28] Senate Report at 25. On March 28, 1986, EPA proposed rules that would require corrective action for releases which extended beyond the facility boundary. The proposed rules would also require facilities to provide financial assurance for corrective action extending beyond the boundary. 51 Fed. Reg. 10717 (Mar. 28, 1986).
[29] HSWA Sec. 232(b), amending RCRA §3008(e)(2)(A) and (B).
[30] HSWA Sec. 232(a), amending RCRA §3008(d)(1).
[31] RCRA §3008(d)(1).
[32] HSWA Sec. 232(a), amending RCRA §3008(d)(3).
[33] HSWA Sec. 232(a), amending RCRA §3008(d)(2)(C).
[34] HSWA Sec. 232(a), amending RCRA §3008(d).
[35] HSWA Sec. 403(d), adding RCRA §2002(d) and HSWA Sec. 403(b), adding RCRA §7012.
[36] 130 Cong. Rec. S9161 (daily ed., July 25, 1984).

[37] *Id.*
[38] *United States v. Johnson & Towers, Jack W. Hopkins and Peter Angel*, 741 F.2d 662, 666 (3rd Cir. 1984).
[39] Senate Report at 5.
[40] HSWA Sec. 245, adding RCRA §3017(a).
[41] 51 Fed. Reg. 8758 (Mar. 13, 1986).
[42] Adding RCRA §3017(a)(2).
[43] Senate Report at 48.
[44] 51 Fed. Reg. 8758 (Mar. 13, 1986).
[45] HSWA Sec. 245, adding RCRA §3017(g).
[46] 40 C.F.R. §262.50(d).
[47] Senate Report at 48.
[48] HSWA Sec. 245, adding RCRA §3008(d)(6).
[49] HSWA Sec. 103(a), adding RCRA §2008(a).
[50] House Report at 62.
[51] Conference Report at 82.
[52] Senate Report at 4.
[53] Conference Report at 109 and 110.
[54] Adding RCRA §3007(e)(1).
[55] House Report at 72.
[56] Senate Report at 41 and 42, House Report at 72.
[57] Adding RCRA §3007(e)(2).
[58] Senate Report at 42.
[59] House Report at 72, Senate Report at 42.
[60] HSWA Sec. 230, adding RCRA §3007(d) (for state facilities); HSWA Sec. 229, adding RCRA §3007(c) (for federal facilities).
[61] Conference Report at 109.
[62] HSWA Sec. 229 and 230, adding RCRA §3007(c) and (d).
[63] HSWA Sec. 244, adding RCRA §3016(a).
[64] HSWA Sec. 244, adding RCRA §3016(b).
[65] Senate Report at 44.

Chapter 17

The Role of the States

> NATIONAL POLICY—The Congress hereby declares it to be the national policy of the United States [to establish] a viable Federal-State partnership to carry out the purposes of this Act and [to ensure] that the Administrator will, in carrying out the provisions of Subtitle C of this Act, give a high priority to assisting and cooperating with States in obtaining full authorization of State programs under Subtitle C.[1]

In enacting RCRA in 1976, Congress intended that states would eventually assume the responsibility of implementation and enforcement of Subtitle C. Following the pattern of other environmental laws, Congress sought to avoid confusing and wasteful duplication of effort by two levels of government. In theory, the partnership concept was rational and simple: EPA, with its technical expertise, would develop the program, and the states, with financial assistance from Washington, would carry it out. In practice, however, the federal-state partnership has been far more complex, just as in the case of other environmental laws.

State Authorization

Under Section 3006(c) the first step in the federal-state relationship begins with the state's application for interim or temporary authorization to run the RCRA program. A state may apply for authorization for one or more components of the Subtitle C program including Sections 3002 (generators), 3003 (transporters), 3004 (treatment, storage and disposal facilities), and 3005 (permits). A state's application for interim authorization must be approved by the EPA Administrator if the state's RCRA program is determined to be "substantially equivalent" to the federal program. To encourage states to obtain final authorization (which requires that the state's program be "equivalent" to EPA's), the interim authorization was made temporary; under the 1976 Act, all interim authorizations expired on January 26, 1985. The expiration of a state's authorization meant that the RCRA program reverted to EPA, and it then was the federal government's responsibility to implement and enforce the program in that state. Historically, federal agencies have been loathe to see environmental programs, once delegated, revert to their jurisdiction, particularly during times of declining federal agency budgets and staff resources.

By September 1982, six years after RCRA's enactment, 34 states had received interim authorization for the generator and transporter components, and five of those had received interim authorization to issue permits. No state, however, had received final authorization for even a single component of treatment, storage, and disposal facilities, primarily because no state could apply for final authorization until after regulations had been issued by EPA, which wasn't until July 26, 1982.* This left a little more than two years for states

*As of May 1985, 45 states and jurisdictions had received interim authorization for generators and transporters and 25 to issue permits for treatment, storage, and disposal facilities as well.

to analyze the EPA rules and adopt equivalent programs, a process that required promulgation of state regulations and, in some states, adoption of new laws by state legislatures, followed by promulgation of regulations. Moreover, there was no certainty that the action by the state agencies and legislatures would succeed in obtaining final EPA authorization. From the states' perspective, EPA's constant revision of its RCRA rules was making it difficult, if not impossible, to adopt programs which were, in fact, equivalent to EPA's most recent set of regulations. The states complained of a constantly moving target, a problem the states have with most of the federal environmental programs.

The "moving target" phenomenon was also a major concern of Congressman Florio, who included a provision in H.R. 2867 designed to solve the problem. The "fix" was to authorize EPA to approve a state program equivalent to the federal program that was in effect one year prior to the submission of a state's application or January 26, 1983, whichever is later.[2] Thus, from the states' perspective, the previously moving target would be frozen. The House bill also called for extending the deadline for states to receive final authorization to January 31, 1986. These changes, which were included in the final version of the legislation,[3] prevented the reversion of about a dozen state programs, and enabled all states to submit a complete application for final authorization.

Despite the fix on the moving target, a state's so-called final authorization, would not really be "final."* It would be final only for the RCRA program in effect prior to the 1984 Amendments. To be authorized for the "new RCRA," a state must follow essentially the same procedure: interim authorization may be granted to states with programs that are "substantially equivalent," and final authorization may be granted once those programs are determined to be "equivalent." Another important change from the 1976 law is the absence of a statutory deadline for the expiration of the interim authorizations. Instead, Congress amended Section 3006(c) to direct the Administrator to promulgate a date by which interim authorizations for the new 1984 provisions would terminate.

Congress also provided that, pending EPA's review of a state's application for interim or final authorization, EPA and the state may enter into an agreement under which the state "may assist in the administration of the requirements and prohibitions" imposed by the 1984 Amendments.[4]

Given that the changes made by the 1984 Amendments obviously would have a major impact on the permit requirements for treatment, storage, and disposal facilities, Congress wanted to avoid any obstacles that would cause

*In its Codification, EPA considered the question of whether states that believe they already have substantially equivalent or equivalent provisions to some of the 1984 Amendment requirements would automatically obtain interim or final authorization for those requirements. 50 FR 28731 (July 15, 1985). EPA rejected the idea, noting that otherwise the public, EPA, and the State Attorney General would be deprived of the opportunity of making an independent judgment. EPA indicated, however, that the application for authorization in this situation need not be lengthy.

further delays in the issuance of permits. One such obstacle would be the obsolescence of the state's permitting authority. In other words, a state that had obtained final authorization to issue permits under the "old RCRA" would not be able to issue permits for any facility subject to a requirement imposed by the 1984 Amendments—even if the requirement was relatively minor. To avoid this problem, the 1984 Amendments direct EPA, pending state authorization, to coordinate with the states on procedures for issuing permits affected by the new law.[5]

Application of the 1984 Amendments to Authorized States

Section 228 of the 1984 Amendments[6] provides that any requirement or prohibition applicable to the generation, transportation, treatment, storage, or disposal of hazardous waste imposed by the new law shall take effect in each authorized state on the same date as in nonauthorized states. Under this provision, EPA is required to carry out the requirements and prohibitions of the 1984 Amendments directly, not only in non-authorized states, but also in a currently authorized state until the state obtains authorization for those new provisions.

This provision represents a significant change: prior to 1984, revisions to Subtitle C did not take effect immediately in authorized states. These states were given time to amend their hazardous waste laws and regulations to come into compliance with the federal requirements (one year for regulatory amendments, two years if legislative enactments were necessary).

Congress' rationale for imposing the prohibitions and requirements on a uniform basis was quite simple. Under the old system the restrictions on land disposal of hazardous wastes would take effect in non-authorized states substantially earlier than they would be imposed in authorized states. The result would have been a "patchwork quilt" effect with, for example, the ban on liquids in landfills enforced by EPA in Wyoming (a non-authorized state) but with continued disposal in Montana (an authorized state). For a period of several years, authorized states would have become the dumping grounds of wastes banned in other states, thus undercutting incentives to develop waste disposal alternatives such as waste reduction, treatment, and recycling.

Underlying these rather modest improvements to the institutional relationships between Washington and the state capitals was a strong belief on the part of many Senate and House members that RCRA could be implemented much more effectively if the petty squabbling over the authorization process could be ended. The cause of the ongoing disputes, according to Congressman Florio, was "entirely too much emphasis by EPA on procedural, even nitpicky matters, rather than the substance of the states' programs."[7]

Florio felt it was necessary to spell out in the statute a congressional declaration that "establishing a viable federal-state partnership" to carry out RCRA was a "National Policy." Florio said that this policy was:

> simply intended to convey Congress' clear and unmistakable message to EPA that it must make much greater effort to assist States in obtaining authorization for RCRA programs. . . . EPA should remember that RCRA is a partnership and that States should be encouraged to develop innovative and more efficient ways of carrying out this statute.[8]

A similar sense of the states' role vis-a-vis the federal government was expressed in a colloquy between Florio and Congressman Hal Daub (R-Nebraska), concerning the process of obtaining authorization:

> Mr. DAUB. My State has urged me to ask for the committee's definition of "equivalency" as utilized in section 3006(b) of the Resource Conservation and Recovery Act.
>
> It is my understanding that this yardstick by which States are judged to be eligible for authorization is currently being interpreted by the Environmental Protection Agency as being synonymous with "identical."
>
> Since little report language on floor debate addressed this important criterion during the 1976 and 1980 consideration of this act, I am interested in the chairman's interpretation.
>
> Mr. FLORIO. Mr. Chairman, the gentleman's inquiry is certainly welcome. I, too, have detected confusion surrounding this word. As you know, regardless of which word Congress might have chosen to gauge a State's program against the Federal program for authorization, a word with no congressional guidance could be misunderstood.
>
> For that reason, I would like to take a few moments to clarify the committee's intent.
>
> States offer a unique set of experiences in the practical administration of this Nation's hazardous waste program, as evidenced by amendments under consideration today. State programs have evolved over the years into more sophisticated programs in response to issues and problems raised by hands-on experience. For example, 20 out of 45 States recently surveyed already regulate small generators more stringently than EPA.
>
> Moreover, California's ban on land disposal of certain chemicals serves as a prototype for this bill.
>
> This realm of experience should be brought to bear in the hazardous waste program under subtitle C of RCRA, rather than removed from its consideration.
>
> Mr. DAUB. I have an additional question: How is the administration to forge unique State program elements with the Federal program?
>
> Mr. FLORIO. State provisions can differ as long as they address the Federal program requirements, and include State re-

> quirements at least as stringent as those of the Federal program. State requirements should be equivalent in overall effect to the Federal program, without the necessity of showing point-by-point equivalence. . . . It was not Congress' intent to require States to xerox the Federal programs.[9]

The Senate Report echoed this concern: "State program requirements and procedures which achieve the same results intended by the requirements of Subtitle C should be deemed 'equivalent.' "[10]

Despite this clear expression of legislative intent, it seems unlikely that EPA's relationship with the states will be dramatically altered as a result of passage of the 1984 Hazardous and Solid Waste Amendments. For one thing, the 1984 Amendments contain explicit instructions from Congress on which wastes to ban from land disposal, how frequently facilities should be inspected, and so forth. While states have been able to establish rules that are more stringent than the federal requirements, it will be difficult in many cases for a state to come up with innovative methods to accomplish the specific objectives Congress mandated.

Thus, at least to a certain extent, the photostatting or cloning of EPA's RCRA rules by states will continue for the foreseeable future. Once states gain more experience in managing the new requirements, they unquestionably will develop different approaches and methods of implementation. This development should not come as a surprise, since it was the innovative efforts of the states, particularly California, upon which many of the reforms in the 1984 RCRA Amendments are based.*

*For an excellent review of innovative hazardous waste programs being implemented by states, see Lennett and Greer, *State Regulation of Hazardous Waste,* 12 Ecology Law Quarterly 183 (1985).

Chapter 17 Footnotes

[1] HSWA Sec. 101(b), amending RCRA §1003(b)(7).

[2] HSWA Sec. 225, amending RCRA §3006(b).

[3] HSWA Sec. 227, amending RCRA §3006(c). The change and other changes to the state authorization program are reflected in EPA regulations at 40 C.F.R. §271.

[4] HSWA Sec. 227, adding RCRA §3006(c)(3).

[5] HSWA Sec. 227, adding RCRA §3006(c)(4).

[6] Adding RCRA §3006(g). Amending RCRA Sec. 3008(e)(1)(b)(ii), now RCRA Sec. 3008(d)(2)(C), 42 U.S.C. §6928(d)(2)(C). On January 6, 1986, EPA proposed new rules to address one of the difficulties that states would encounter in attempting to achieve final authorization for the multitude of requirements imposed by the 1984 Amendments. 51 Fed. Reg. 496 *et seq.* (Jan. 6, 1986). Under the proposal, EPA would "cluster" most of the requirements as a group of regulations promulgated during a three-year period ending July 1, 1987. States would have one year from that date (two years if they need statutory changes) to come into compliance with the federal regulatory requirements. After the three-year cluster period, the authorization schedule would switch to a one-year "cluster." Under this scheme, states would have one year from each July 1st (two years in the case of a statutory change) to bring their requirements into conformance with any RCRA regulations promulgated by EPA during the previous 12 months.

[7] 129 Cong. Rec. H9154 (daily ed., Nov. 3, 1983).

[8] *Id.*

[9] 129 Cong. Rec. H6507 (daily ed., Aug. 4, 1983).

[10] Senate Report at 62.

Chapter 18

Groundwater Monitoring and Financial Responsibility: Certification of Compliance

During 1983, a series of reports by the General Accounting Office[1] and by EPA[2] indicated massive and widespread noncompliance with the groundwater monitoring rules that EPA had issued in November 1981.[3] The rules, which applied to all hazardous waste landfills and surface impoundments, were designed to be the first line of defense in the event a disposal facility began to leak. The rules were premised on the belief that detection of a leak in the early stages of groundwater contamination could prevent much more serious problems later on.

The regulations EPA issued in 1981 require each facility to have a sufficient number of downgradient wells (EPA originally stated that three would be sufficient for most facilities) which would be tested periodically to determine whether hazardous constituents exceeded the "background level." The "background level" could be determined by measuring the water quality at the facility's one upgradient well. To be effective in testing for leaks, the wells had to be placed where they would be likely to detect the presence of hazardous constituents, and they had to be sampled on a regular basis (every three months according to EPA's regulations).

Despite the obvious utility of an effective groundwater monitoring program, the vast majority of facility owners and operators either ignored the requirements or made inadequate efforts to come into compliance with them. According to the General Accounting Office, in a report requested by Congressman Florio, the average noncompliance rate for Illinois and North Carolina was 78 percent.[4] GAO reviewed the records of two other states, California and Massachusetts, and found that not enough inspections had been completed to determine a compliance rate.[5] A survey conducted by EPA[6] in 1983 covering 586 facilities showed that only 15 percent were in full compliance with the groundwater monitoring requirements. In addition, 148 of these facilities in the survey which *had* implemented groundwater monitoring programs failed nonetheless to satisfy certain basic requirements. For example, of these 148 facilities:

- 24 percent did not have adequate upgradient wells.
- 32 percent did not have adequate downgradient wells.
- 25 percent had problems related to sampling and analysis procedures.
- 36 percent did not maintain required records.
- 40 percent did not submit required reports.

Reacting to what appeared to be a flagrant scofflaw attitude on the part of facility owners and operators—as well as a lackadaisical enforcement policy by EPA and the authorized states—Congress enacted another "hammer" provision. On October 6, 1983, the House passed an amendment offered by Congressmen Florio, Eckart, and Lent. That amendment, among other things, required the owner/operator of an interim status land disposal facility to certify compliance with all applicable groundwater monitoring and financial responsibility requirements.[7] The owner/operator also was required to submit a completed "Part B" application for a final permit. The certification had to be completed within one year of enactment; failure to certify would automatically terminate the interim status designation of the facility. Not only would this provision require the closure of the facility, but any disposal of hazardous waste beyond the termination date would constitute a criminal violation under Section 3008(d)(2)(A)—a violation punishable by five years imprisonment and a $50,000 fine per day of violation.

The House adopted the amendment without debate. The most succinct reference to it came from Congressman Edwin Forsythe (R-New Jersey), who stated that, because the groundwater monitoring requirements had been in effect since November 1981, "there is no excuse for noncompliance at this late date."[8]

Although the Senate bill did not contain a similar requirement, the Conference Committee adopted the House provision without discussion, enacting it into law as RCRA Section 3005(e)(2).[9] In fact, the only in-depth public discussion of the provision occurred at an April 29, 1985, hearing by the Oversight and Investigations Subcommittee of the House Energy and Commerce Committee. There, it was disclosed that the Subcommittee staff had conducted a comprehensive survey which had revealed that only 41 percent of the facilities had "barely adequate" compliance with EPA's groundwater monitoring requirements. Additionally, many facility owners had installed no wells, and others with installed wells had not conducted the necessary sampling and analysis.

EPA's original response to the hammer provision for groundwater monitoring enacted in the Amendments was an attempt to construe the term "compliance" broadly. Under EPA's proposal, facilities would not have to be in actual compliance by the November 8, 1985, deadline so long as they were on an EPA-approved "compliance schedule." This flexible deadline approach (which has been referred to as "spiritual compliance") turned out to be a short-lived policy. After a brief storm of protest at the time of the subcommittee hearings, EPA changed course, stating that it would demand "complete physical compliance" with the groundwater monitoring requirements. Despite that assertion, however, the groundwater monitoring policy still was in a state of flux in the fall of 1985. In its Codification Rule, EPA eased the burdens of the certification requirement somewhat by interpreting it to affect *only the unit or units* at the hazardous waste management facility for which the information is not submitted.[11] Thus, if a facility had an incinerator and land disposal unit and submitted a certification only for the incinerator, interim status would terminate for the land disposal unit, but not the incinerator.

The estimates of how many interim status land disposal facilities would be forced out of business varied widely, but EPA Administrator Lee M. Thomas in August 1985 predicted that 40 percent of the nation's 1,575 land disposal facilities would close as a result of their inability to comply.[10] As it turned out, approximately two-thirds of the Nation's interim status land disposal facilities lost their right to continue operations. According to EPA, only 521 facilities had submitted Part B applications and certified to compliance with the groundwater monitoring and financial responsibility requirements by the November 8 ("black Friday") deadline.

At the time of enactment, the requirement for certifying financial responsibility was not considered unrealistic. Subsequently, however, the insurance market for hazardous waste facilities practically collapsed. As a result, EPA announced it would consider alternatives to certifying compliance with the financial responsibility requirements.[12] But the November 8, 1985, deadline passed and EPA did not modify or suspend any of those requirements. The congressional reaction was somewhat more sympathetic. On December 16, 1985, the House passed a bill (H.R. 3917) authored by Congressman James T. Broyhill which provided an opportunity for land disposal facilities which had submitted their Part B applications and were in compliance with the groundwater monitoring requirements, but had been unable, despite good faith efforts to obtain liability insurance, to retain interim status until November 8, 1986. The bill would also direct EPA to revise the financial responsibility requirements in 40 C.F.R. Section 265.147 and make it easier for facility owners to come into compliance.

Curiously, the interest group whose voice had been silent during the entire congressional debate on the certification hammer was the owners and operators themselves. Clearly a sizable number could afford to ignore the requirements because they never intended to remain in operation if additional regulatory burdens were imposed. Other facilities presumably remained out of compliance because the owners and operators had trouble understanding what was required. Others simply remained ignorant of the law, and they may continue to carry on their operations as usual until enforcement action is taken.

Chapter 18 Footnotes

[1] GAO, Interim Report on Inspection, Enforcement and Permitting Activities at Hazardous Waste Facilities (Sept. 21, 1983).

[2] EPA, *Evaluation of the Ground Water Monitoring, Interim Status: Phase II Report* (March 10, 1983).

[3] *Id.* at 3; 40 C.F.R. §265, Subpart F.

[4] GAO Interim Report, supra n.1 at 6.

[5] *Id.* at 7.

[6] EPA Ground Water Report, supra n.2 (information cited in GAO report, supra n.1, at 9).

[7] The certification requirements are set forth at 40 C.F.R. Part 265, Subpart F (groundwater monitoring) and H (financial responsibility).

[8] 129 Cong. Rec. H8142 (Oct. 6, 1983); Congressman Lent believed that provision would "assist EPA in performing its responsibilities since owners will now be forced to close down if they fail to certify . . ." 129 Cong. Rec. H8135 (Oct. 6, 1983).

[9] HSWA Sec. 213.

[10] Hazardous Waste Report at 4 (Sept. 2, 1985).

[11] 50 Fed. Reg. 28727 (July 15, 1985).

[12] 50 Fed. Reg. 33902 (Aug. 21, 1985).

Chapter 19

National Groundwater Commission

Summary

A Groundwater Commission is to be Established and:

- By November 8, 1985, it shall complete and submit to Congress and the President its preliminary report which includes findings and conclusions.
- By October 30, 1986, it shall complete and submit its final report which includes findings and conclusions.

The Study Must Assess Numerous Specific Items, Including:

- The amount, location, and quality of the nation's groundwater.
- The relationship between groundwater contamination and groundwater withdrawal.
- The relationship between surface water pollution and groundwater pollution.
- The effects of land-use patterns on groundwater.
- The methods of abating groundwater contamination and their costs and benefits.
- The adequacy of existing law for protecting groundwater.
- The effects of underground injection practices on groundwater.
- The adequacy of existing groundwater research.

Commission Shall be Composed of 19 Members to be Appointed as Follows:

- Six by the Speaker of the House from among Members of the House, with the Speaker also appointing the Commission Chairman from the House members.
- Four by the Majority Leader of the Senate from among Members of the Senate.
- Eight by the President.
- The Director of Congress' Office of Technology Assessment.

Commission Funding and Life:

- Authorized up to $7 million for fiscal years 1985-87.
- Expires on January 1, 1987.

> There is a growing belief that the next domestic crisis we may face as a Nation is a water crisis, and that its solution may be more expensive and more elusive than the energy crisis with which we have been struggling over the past decade. If we are to avert such a crisis, we must move now to provide effective protection of our irreplaceable groundwater resources.[1]

Politicians on Capitol Hill are always warning of a major impending "crisis," which they can resolve only by taking bold, immediate, and decisive action. Was there reason to believe that the groundwater crisis was real? If so, would Congress enact legislation to resolve the crisis?

Although the breadth and scope of the 1984 Hazardous and Solid Waste Amendments offer ample proof that the congressional authors were not afraid to tackle complex environmental problems, the task of developing a truly comprehensive solution to groundwater contamination was never seriously considered. Because the complexity of the problem defied a quick legislative fix, Congress resorted to a tactic it frequently reserves for postponing the need to resolve differences: creation of a congressional commission to study the issue and report back with recommendations on what to do about it.*

In reviewing American society's gradual awareness of the groundwater contamination issue, one recalls the "out of sight, out of mind" cliche. Groundwater is a resource which long has been overlooked and taken for granted. Yet, by volume alone, groundwater constitutes more than 96 percent of the fresh water in the United States.[2] In its "Second National Water Assessment, The Nation's Water Resources, 1975-2000," the U.S. Water Resources Council estimated that between 33 quadrillion and 59 quadrillion gallons of groundwater—enough to supply current withdrawals for 200 to 300 years, even with no recharge of aquifers—are located within 2,500 feet of the surface.[3]

Citing U.S. Geological Survey data, the Congress' Office of Technology Assessment (OTA) has estimated that groundwater is relied on for about 50 percent of drinking water supplies.[4] In a report issued in 1984,[5] OTA stated groundwater is used to supply water for nearly 80 percent of rural domestic and livestock needs, about 40 percent of irrigation needs, and nearly 25 percent of self-supplied industrial needs, other than thermoelectric power. Groundwater is also crucial in maintaining stream flows and in avoiding salt-water intrusion into fresh water.[6]

Reliance on groundwater supplies varies from region to region throughout the U.S. In the Great Lakes region, for instance, groundwater provides about 11 percent of public water supplies, but in the Rio Grande region it provides nearly three-quarters of the public water supplies. In the Upper Colorado River Basin, groundwater supplies about 12 percent and 1 percent respectively for

*In addition to voting to establish a Groundwater Commission, Congress in the 1984 Amendments established several other programs aimed primarily at the groundwater problem, among them a program to address leaks from underground storage tank restrictions on land disposal, and minimum technological requirements for land disposal facilities.

rural uses and for irrigation, while the comparable figures are 100 percent in New England for rural uses and 90 percent in the Upper Mississippi for irrigation. Some places are entirely dependent on groundwater for their primary drinking water supplies, and Phoenix, Arizona, for instance is the largest city in the U.S. in that category.[7]

As a percentage of total supply, relatively little groundwater is contaminated. It generally is estimated that "only" 1 to 2 percent of groundwater is contaminated by pollution.[8] The figure, however, may well be misleading, and in any event it represents an enormous quantity by volume. The important point is that in many cases groundwater has become polluted precisely where it is most needed. Equally disturbing is the fact that groundwater contamination is increasing. As OTA said in its October 1984 report to Congress, *Protecting the Nation's Groundwater From Contamination*, "There is a growing consensus that the quality of groundwater is in decline. Incidents of contamination are being reported with increasing frequency and have now occurred in every state."[9]

Moreover, groundwater contamination, by its very nature, is inaccessible and difficult to clean up. The Conservation Foundation noted in its "State of the Environment 1982" report:

> One area may be contaminated and an area only 100 feet away may not. As pollutants travel from land surface to aquifer, soil may act as a natural filter. Once in the ground water, however, contamination is likely to stay for many years. It is usually impractical, if not impossible, to remove contaminants from water that is still in the ground.[10]

Noting that some purification technologies are "prohibitively expensive," the Conservation Foundation report says: "Removing salts or toxics, for example, usually requires expensive purification equipment that is probably not within the economic reach of most municipalities that depend on groundwater, much less the large number of individuals who rely on private wells."[11]

In addition to the technical and financial difficulties inherent in correcting existing groundwater contamination, there are substantial administrative and institutional challenges to be overcome in addressing groundwater pollution on a broad scale. As Congress took note of the increasing number of reports on groundwater contamination,[12] it became evident that "something had to be done." If the source of the groundwater contamination could be attributed *just* to hazardous waste dump sites or leaking underground storage tanks, Congress may well have been content to rely solely on the stringent new reforms being imposed by the 1984 Amendments.

However, the known sources of groundwater pollution simply did not fall neatly into one or two clearly identifiable and manageable categories. Instead, numerous segments of American society were found to be contributing to the groundwater contamination problem. Congress wanted to address the problems created not "only" by waste sites and underground tanks but a range of sources such as agricultural applications of pesticides, herbicides, and fertilizers; runoff from salts used to melt ice on roads and highways; septic tanks; accidental spills of chemicals during transportation; pipeline ruptures; brines from petroleum exploration; and radionuclides from commercial reactors, defense facilities, hospitals, as well as natural sources.[13]

Whether "can-do" optimism is a curse or a blessing, Members of Congress frequently have it. No matter how daunting or complicated the challenge, the solutions often can be found in a "comprehensive, coordinated policy." According to the 98th Congress, even the groundwater contamination problem eventually would have to yield in the face of such an effort.

Thus, on November 3, 1983, the House of Representatives passed an amendment to H.R. 2867 to establish a National Groundwater Commission. According to James J. Florio, the co-author* of the Groundwater Commission amendment, "While we have taken every possible legislative action in this bill to guard against groundwater contamination [from hazardous wastes], I believe it is imperative that we take the further step of development of a comprehensive national plan to insure that everything that can be done will be done and that the Nation's supply of water will be safe and pure."[14]

New York Congressman Norman Lent, Republican floor manager of H.R. 2867, also supported the amendment:

> While I believe that the states should ultimately have the primary responsiblity for protecting groundwater, I favor a coordinated effort at the national level to determine what is needed to prevent further deterioration of our groundwater supplies. . . . [A] National Groundwater Commission will provide us with major progress toward establishing a comprehensive national groundwater policy.[15]

To Congressman Guy Molinari (R-New York), the idea of such a commission was "long overdue." In urging his colleagues to establish the commission, Molinari had this to say:

> We know that our groundwater resources are threatened with contamination . . . yet we lack comprehensive information on the nature and scope of these threats. Six federal statutes, many of which overlap, govern aspects of groundwater problems. Additionally, state protection measures are inconsistent, fragmentary, and limited. A Commission will transcend this jurisdictional confusion and objectively determine the best remedial and preventive action to take.[16]

To Congressman George Brown (D-California), another advocate of establishing the commission, the groundwater issue clearly transcended state borders and political jurisdictions, a key justification for a federal approach. "One of the loopholes in RCRA is that there is no national policy to control

*Congressman Robert Edgar (D-Pennsylvania), co-authored the Groundwater Commission amendment, and was the original sponsor of a similar provision in the previous Congress in H.R. 6307, the earlier RCRA reauthorization bill.

contamination of groundwater," Brown said. "As a result there is no uniform approach—and precious little cooperation between the federal government and the states—in addressing one of the most dangerous threats to public health." Brown emphasized that "aquifers do not recognize state boundaries, which is all the more reason for a policy that is national in scope."[17]

The Groundwater Commission Amendment, facing no House opposition, was adopted on a voice vote. The Senate RCRA reauthorization bill, however, did not contain a similar provision, and the Senate conferees were not nearly so enthusiastic as their House counterparts about the possible accomplishments of such a commission. While not endorsing establishment of a Groundwater Commission, the Senate conferees did not see the issue as meriting their opposition. They only reluctantly accepted the Groundwater Commission Amendment as part of a final compromise package. The House and Senate conferees also agreed to a $7 million authorization for the Commission.

Congress' Mandate to Groundwater Commission

As it had been throughout much of the 1984 Hazardous and Solid Waste Amendments, Congress was specific in its mandate to the National Groundwater Commission it sought to establish under Section 704. The Groundwater Commission amendment, as approved by the House bill and accepted by House and Senate conferees as part of an overall compromise, calls for the Commission to undertake 21 specific duties.

Even if the Groundwater Commission itself never comes to fruition, it is instructive to review the sweeping mandate for action that Congress had in mind in requiring the groundwater study. That mandate provides insights into the kinds of issues that may interest legislators as they review groundwater issues in the future. It requires specifically that the Commission:

"1) Assess generally the amount, location, and quality of the Nation's groundwater resources.

2) Identify generally the sources, extent, and types of groundwater contamination.

3) Assess the scope and nature of the relationship between groundwater contamination and groundwater withdrawal and develop projections of available, usable groundwater in future years on a nationwide basis.

4) Assess the relationship between surface water pollution and groundwater pollution.

5) Assess the need for a policy to protect groundwater from degradation caused by contamination.

6) Assess generally the extent of overdrafting of groundwater resources, and the adequacy of existing mechanisms for preventing such overdrafting.

7) Assess generally the engineering and technological capability to recharge aquifers.

8) Assess the adequacy of the present understanding of groundwater

recharge zones and sole source aquifers and assess the adequacy of knowledge regarding the interrelationships of designated aquifers and recharge zones.

9) Assess the role of land-use patterns as these relate to protecting groundwater contamination.

10) Assess methods for remedial abatement of groundwater contamination as well as the costs and benefits of cleaning up polluted groundwater and compare cleanup costs to the costs of substitute water supply methods.

11) Investigate policies and actions taken by foreign governments to protect groundwater from contamination.

12) Assess the use and effectiveness of existing interstate compacts to address groundwater protection from contamination.

13) Analyze existing legal rights and remedies regarding contamination of groundwater.

14) Assess the adequacy of existing standards for groundwater quality under State and Federal law.

15) Assess monitoring methodologies of the States and the Federal Government to achieve the level of protection of the resource as required by State and Federal law.

16) Assess the relationship between groundwater flow systems (and associated recharge areas) and the control of sources of contamination.

17) Assess the role of underground injection practices as a means of disposing of waste fluids while protecting groundwater from contamination.

18) Assess methods for abatement and containment of groundwater contamination and for aquifer restoration including the costs and benefits of alternatives to abatement and containment.

19) Assess State and Federal groundwater law and mechanisms with which to manage the quality of the groundwater resource.

20) Assess the adequacy of existing groundwater research and determine future groundwater research needs.

21) Assess the roles of State, local, and Federal Governments in managing groundwater quality.

Makeup of the National Commission

In designing the Groundwater Commission, Congress went to extraordinary lengths in detailing how Commission members would be chosen. Congress specified that six members of the House of Representatives would be appointed by the Speaker of the House to serve on the commission. But it went further than that. It said two of those House members must be members of the Energy and Commerce Committee, two of the Public Works and Transportation Committee, and two of the Interior and Insular Affairs Committee, thereby embracing the three House committees with jurisdiction over laws most directly related to groundwater.[18]

Contrast the House appointee provision with the one for choosing senators. Here, there was an uncharacteristic willingness on the part of the "upper chamber" to settle for less than numerical equality with the House. Thus there are only four Senate members to the House's six. The lopsided ratio reflects the Senate's genuinely lower level of interest in the Groundwater Commission in the first place. The Senate left it to the good judgment of the Senate Majority Leader to pick from among senators interested in serving on the commission, not specifying particular committee assignments that must be represented on the commission. While the senators eventually named to the commission almost unquestionably would come from the Senate committees with jurisdiction over groundwater related laws, there is something of a "We just don't care" attitude in the Senate's more casual approach to the naming of commission members.

With the Senate and House now providing 10 of the commission's 19 members, how do the remaining nine members get named to this congressional commission? Once again, Congress was unusually specific in answering that question.

It specified that the Director of the congressional Office of Technology Assessment be a member and left the appointments of the eight remaining Commission members to the President. But, given the Democratic House's keen interest in just how the commissioners would be named, it did not leave the Republican President a free hand in selecting commissioners.

Instead, the Amendments direct that four of the presidentially-appointed members be chosen from among a list of nominations submitted by the National Governors Association. Careful to avoid the perennial East/West split when it comes to water resources issues, Congress mandated that the President select from the Governors' list two persons representing groundwater appropriation states and two representing groundwater riparian states.

Congress also required that the President select a candidate from among a joint list submitted by the National League of Cities and the U.S. Conference of Mayors, and one from among a list submitted by the National Academy of Sciences. The statute stops short of legislating outright harmony among the League of Cities and the Conference of Mayors in the event they cannot quite agree on a single list.

Among other lists from which the President is to choose is one to be submitted by unspecified "groups, organizations, or associations of industries the activities of which may affect ground water." It might seem that such a compilation could result from the commission's end product, but that is not what Congress had in mind here. While the statute makes no attempt to narrow which particular groups might submit nominees, it should be interesting to watch which ones make the effort to indicate that they *do* affect groundwater and which ones avoid the issue. It is possible that the President could find himself either inundated or high and dry when it comes to reviewing submittals from this undefined collection of interests that affect groundwater.

The President's final appointee to the Groundwater Commission is to be the traditional "public member." In this case, a list of nominations is to come from "groups, organizations, or associations of citizens which are representative of persons concerned with pollution and environmental issues. . . . " In addition, the groups submitting the list (again it is supposed to be a single list representing a consensus and not scores of individual lists) are themselves to have "participated" at the state or federal level (locals need not apply apparently) in studies, administrative proceedings, and/or litigation relating to groundwater.

Despite all the detailed specifications as to what the Groundwater Commission is to study and how its members are to be chosen, it appears entirely possible that the Commission may never be formed. More than a year after the 1984 Amendments were enacted—and beyond the time that the would-be commission's preliminary findings and conclusions were to be submitted to Congress—neither the Administration nor the Congress had made any move to name members.

While Congress had the foresight to mandate an early reporting deadline, it did not specify a suitably early deadline for its and the Administration's appointment of the commissioners, who in turn would appoint the staff to get the necessary studies completed. Delays in appointing members to past commissions, such as the National Commission on Air Quality, have led to problems in their fulfilling their legislative mandates.

Given the lukewarm attitude of the Senate, the constraints placed on the Republican President in selecting members, and the budgetary anxieties of both the Congress and the Executive Branch in the face of enormous federal deficits, the few Democratic House members actively interested in advancing the Groundwater Commission may not succeed in launching it.

Outlook for the Commission . . . Enormous Challenges

In addition to the real possibility that the Groundwater Commission members never will be appointed, the Commission also faces the difficulty of having a final reporting deadline of October 30, 1986,[19] and an expiration date of January 1, 1987.[20] If somehow the Commission can be launched and its deadlines extended, the precedents for congressional study commissions in the environmental field still do not augur well for the Commission's final report and conclusions becoming a "mandate for action."

Under the 1972 Federal Water Pollution Control Act Amendments, Congress established the National Commission on Water Quality to study and report back on future legislative options for achieving "fishable and swimmable" waters by 1983, as required by law.[21] The Commission, concentrating on issues most in Congress' mind in 1972 when it was established, focused its research primarily on the conventional water pollutants. But by the time the Commission's final report and recommendations were submitted to Congress in 1974, congressional interest had shifted to another concern. When in 1977 Congress amended the Clean Water Act, toxic water pollutants, an issue scarcely addressed by the Water Commission, were the top legislative priority. In large measure, the mere passage of time—between the establishment of the Commission, the submission of its report, and congressional action—had changed the focus of congressional interest. Thus, the Water Quality Commission report suffered a fate not unknown to similar reports—it was overtaken by more immediate issues.

In the air quality program, the history of congressional commissions does little to instill confidence that commissions can blaze the path for subsequent legislative activities. Congress in the Clean Air Act Amendments of 1977 established the National Commission on Air Quality.[22] Facing no statutory deadline for appointing the Air Quality Commission's nine "public members," the Carter Administration delayed nearly a year after enactment before making any appointments. What was to have been a three-year examination had *de facto* truncated by one-third.

Once under way, the Air Quality Commission's performance of its studies was further complicated by the constantly changing makeup of its congressional members. Chairman Robert T. Stafford (R-Vermont) of the Environment and Public Works Committee was nominally a member of the Air Commission, but, from the start, he took the position that he would not participate in its studies for fear of later having his "hands tied" when it came time to consider Air Act amendments in the Senate. In addition, there was a succession of House members, resulting from decisions not to seek re-election or, as in the case of Congressman David Stockman, a decision to join the Reagan Administration. Even those congressional members who were Air Quality Commission members in 1981, when the final report and recommendations were filed, felt only marginally committed to the work product. The sole exception was Commission Chairman Gary Hart (D-Colorado). However, he could not, on his own, successfully carry the Commission's message to an uninterested Congress and soon gave up trying.

As is evident from these examples, the Groundwater Commission will confront an enormous work load and sizeable administrative and political hurdles. Because of these difficulties, particularly the brief time allotted for the Commission's effort, it is unlikely that a commission could provide major "new" insights into the issue of groundwater contamination. Assuming the commission is eventually established, perhaps a more likely result would be the compilation of existing information in a format useful to the Congress.

Chapter 19 Footnotes

[1] House Committee on Government Operations, Rept. No. 1136, 98th Cong., 2nd Sess. 1 (1984).

[2] Weimar, R.A., "Prevent Ground Water Contamination Before It's Too Late," *Water and Wastes Engineering*, Feb. 1980, 30-33.

[3] U.S. Water Resources Council, *The Nation's Water Resources, 1975-2000, Vol. 1: Summary. Second National Water Assessment*, 20.

[4] Office of Technology Assessment, "Protecting the Nation's Groundwater from Contamination, Volume 1," OTA-0-233, October 1984, p.19.

[5] *Id.*

[6] *Id.*

[7] *Id.*

[8] Lehr, J.H. "How Much Ground Water Have We Already Polluted?," *Ground Water Monitoring Review*, Winter 1982, 4-5.

[9] *Protecting the Nation's Groundwater From Contamination, Volume 1*, 21.

[10] *State of the Environment 1982*, The Conservation Foundation, Washington, D.C., 113.

[11] *Id.*

[12] See for example, Congressional Research Service, Groundwater Contamination by Toxic Substances: A Digest of Reports, S.Rpt. 98-131, 98th Cong. 1st Sess., November 1983.

[13] Congressional Research Service, *Groundwater Contamination by Toxic Substances: A Digest of Reports*, S.Rpt. No. 131, 98th Cong. 1st Sess. (Nov. 1983).

[14] 129 Cong. Rec. H9164 (daily ed., Nov. 3, 1983).

[15] 129 Cong. Rec. H9164-65 (daily ed., Nov. 3, 1983).

[16] *Id.*

[17] *Id.*

[18] HSWA Sec. 702, amending RCRA §8002.

[19] HSWA Sec. 704(f)(1).

[20] HSWA Sec. 704(g).

[21] Federal Water Pollution Control Act Amendments of 1972 (P.L. 92- 500), §315.

[22] Clean Air Act Amendments of 1977 (P.L. 95-95), §323.

Chapter 20

Solid Waste/Subtitle D Amendments

Summary

EPA Guidelines and State Obligations:

- By November 8, 1987, EPA must conduct a study to determine if the existing guidelines for solid waste facilities are adequate to protect human health and the environment.
- By March 31, 1988, EPA must promulgate revisions to guidelines for facilities that may receive hazardous household wastes or hazardous wastes from small quantity generators. Guidelines must include requirements for groundwater monitoring and corrective action.
- By April 8, 1989, states must adopt a permit program or other system of prior approval to assure that each solid waste facility that may receive hazardous waste complies with EPA's revised guidelines.

Dioxins from Resource Recovery Facilities:

- EPA must prepare a report on dioxin emissions from resource recovery facilities.
- The report must assess health risks and provide information on methods to control the emissions.
- Based on the report, EPA may publish guidelines on control of dioxin emissions from resource recovery facilities.

Other Important Provisions:

- The states, in determining the size of resource recovery facilities in Subtitle D plans, must consider anticipated needs for recycling projects and also demand for waste-to-energy conversion.
- EPA is authorized to conduct demonstration projects for extending the useful life of sanitary landfills and must submit reports on results. The first report is due by October 1, 1986.

On June 30, 1976, Senator Jennings Randolph, the West Virginia Democrat who chaired the Senate Environment and Public Works Committee, described the scope of federal involvement in addressing solid waste issues. Randolph, who was floor manager of the bill that was to become the Resource Conservation and Recovery Act, had this to say:

> This is not an area which lends itself to extensive planning and operation from the Federal level. The role of the Federal Government in solid waste activities should be one primarily of providing financial and technical assistance. Guidelines authorized by this legislation are to be descriptive of the options available to local and regional bodies responsible for solid waste management. These guidelines should not be taken as directives.[1]

Eight years later Senator Randolph's perspective had changed. "If we give inadequate attention to these general solid waste facilities," he warned his colleagues, "it will only serve to create additional Superfund sites."[2]

He pointed out that "these facilities are recipients of unknown quantities of defined hazardous waste through household waste, unregulated small quantity generator waste disposal, and illegal dumping . . . A high proportion of sites listed on the National Priority List under the Superfund are municipal landfills."[3] The concept of guidelines, in the Senator's view, had to be replaced by a new and more exacting policy. Accordingly, "adequate enforcement and monitoring is imperative to assure that these facilities are not allowed to operate if they are substandard."[4]

In Randolph's opinion, preventing the transformation of municipal dumps, more politely referred to as "sanitary landfills," into Superfund sites required not just active federal leadership, but also the development of stringent regulations for solid waste facilities that the states would be obliged to follow. Consequently, Randolph proposed a sweeping revision of Subtitle D, the solid waste component of RCRA, along with a dramatically expanded effort by EPA.

EPA Guidelines and State Obligations

Randolph's proposal, adopted by the Environment and Public Works Committee, then modified substantially by his own amendment on the Senate floor, directed EPA to:

- conduct a study of the extent to which the existing Subtitle D guidelines are adequate to protect human health and the environment from groundwater contamination. The report, which is to include recommendations for any additional enforcement authorities, is due by November 8, 1987.[5]
- promulgate revisions, by March 31, 1988, of the Subtitle D guidelines for "facilities that may receive hazardous household wastes or hazardous wastes from small quantity generators."* The revisions must include requirements

*A preliminary survey by EPA revealed that at least 37,000 solid waste facilities may be receiving

for groundwater monitoring, criteria for the "acceptable location" of new or existing facilities, and requirements for corrective action.[6]

Parallel obligations are imposed on the states. By April 8, 1989, each state must adopt and implement a permit program or other system of prior approval in order to "assure that each solid waste management facility which may receive hazardous waste will comply with EPA's revised regulatory criteria. EPA is directed to determine whether each state has developed an adequate program and, if not, EPA is authorized to take enforcement action including closing the facilities not meeting the revised criteria.[7]

To assist states in complying with these changes, Congress authorized $75 million through Fiscal year 1988 as grants to states. This total is in addition to the $102 million that Congress authorized for other Subtitle D programs.[8]

Dioxins from Resource Recovery Facilities

One environmental issue which had gained extraordinary prominence during the early 1980s, and which resulted in widespread anxiety, centered on a toxic pollutant that the public came to know as "dioxin."* The concern was fueled by several highly publicized public health incidents—such as the spreading of dioxin-contaminated waste oils on dirt roads in Times Beach, Missouri, which led to an evacuation and federal buyout of the town's residences, and dioxin contamination in the Ironbound section of Newark, New Jersey. The issue was also in the headlines because of the lawsuits brought by thousands of Vietnam veterans against the federal government over health effects they said were related to their wartime exposure to dioxin-contaminated Agent Orange, a powerful defoliant.

One of many aspects of the dioxin issue involved burning solid waste at resource recovery facilities. For example, in July 1979 the Environmental Protection Agency conducted tests on the stack emissions from the Hempstead Resource Recovery Corporation facility in Nassau County, New York. The agency's four samples detected dioxin concentrations of one to nine parts per trillion in the stack emissions.[9]

Despite the fact that this amount of dioxin was extraordinarily small,

hazardous waste or small quantity generator waste. This figure includes 9,283 municipal landfills, 261 industrial landfills, 312 demolition debris landfills, 25,419 surface impoundments, and 1,647 land application units. *Hazardous Waste News,* April 14, 1986, at 146.

*The word "dioxin" is used to refer to a family of chemicals called "dibenzo-p-dioxin." With the addition of various chlorine atoms to this substance, a series of compounds called polychlorinated dibenzo-p-dioxins (PCDD) is formed. There are some 75 distinct types of these PCDDs, the most thoroughly studied of which are those containing four chlorine atoms and formally called the tetrachlorodibenzo-p-dioxins, or TCDDs. Within the TCDDs, there are 22 different isomers or kinds, distinguished from each other specifically in the location at which the chlorine atoms are attached to the dibenzo-p-dioxin molecule. The most toxic of the 22 TCDDs is designated as 2,3,7,8-TCDD, and it is that specific chemical byproduct that is referred to generally as "dioxin" even when it may constitute just a fraction of the PCDDs present.

the Hempstead plant was immediately closed. The public outcry was too intense. A major factor in the closing was the lack of an official standard establishing the emission level of dioxins that was safe.

With the Hempstead facility serving as an example of the volatile nature of the dioxin issue, Congressman Guy V. Molinari, Republican of New York, proposed an amendment that would have directed EPA to adopt regulations "as promptly as practicable" to apply the best available control technology for dioxin emissions for emissions from resource recovery facilities. These standards would apply until dioxin emission standards were expressly adopted under the Clean Air Act.

Molinari pointed out that "[w]ithin the United States, there are . . . 97 [resource recovery] facilities across a span of 36 states. Thirty states have 104 planned resource recovery plants."[10] Molinari urged the House not to allow construction of new plants "in the absence of a standard."[11] "We may discover later," he warned, "when a standard is finally issued, that many operating resource recovery plants are in violation of that standard and emitting dangerous levels of dioxin."[12]

Although sympathetic to the basic objective of the Molinari amendment, the conferees were worried that the term "best available control technology" (BACT) would be difficult to define, and that there could be significant delays in promulgating regulations of this type. In the interim, there could be a virtual halt to development of resource recovery facilities, since there would be considerable reluctance to design or finance a facility until EPA's BACT rules were promulgated.[13] Consequently, the provision requiring the development of BACT regulations was dropped. Instead EPA was directed to prepare a report which would summarize current data on dioxin emissions from resource recovery operations and assess any significant health risks associated with those emissions.[14] The report must also provide information on operating practices which could be used to control dioxin emissions.

Section 102 of the 1984 Amendments authorizes the EPA Administrator to publish advisories or guidelines based on this report, concerning control of dioxin emissions. The Amendments specifically say that they do not preclude or preempt the agency from proceeding independently to regulate such emissions under the Clean Air Act.

While the compromise Conference Committee language represents a significant easing of the House's more aggressive approach to dioxin emissions from resource recovery facilities, Representative Norman Lent (R-New York), the ranking member on the House Energy and Commerce Subcommittee on Commerce, Transportation and Tourism, considered it a victory. "I strongly endorse this provision because there is a resource recovery facility on Long Island which was forced to close by public pressure over dioxin emissions," he said on October 3, 1984, in a floor statement endorsing passage of the Conference Committee bill. "If EPA had dioxin emission regulations in place, this closing could have been avoided." Lent expressed hope that the EPA report and possible subsequent action might lead to the reopening of the Hempstead facility, and might prevent closings of comparable facilities elsewhere.

Size of Waste-to-Energy Facilities

Prior to the 1984 Amendments, Subtitle D of the Resource Conservation and Recovery Act required that states adopt comprehensive solid waste plans. The plans, to be adopted after consultation among state and local officials, were to identify those factors, such as geographic conditions and markets, that would be conducive to carrying out regional solid waste management programs. Congress' basic objective was to foster development of environmentally sound solid waste disposal methods and to encourage resource conservation.

In the interim between passage of the original Subtitle D and enactment of the 1984 Amendments, it became evident that newly built waste-to-energy facilities had an immense appetite for discarded newspapers. Those newspapers, because of their high btu value, were a cheap source of fuel. But they were also important to paper recycling projects. In an attempt to resolve the issue, the House and Senate bills contained identical provisions which specified that, in determining the size of waste-to-energy facilities, comprehensive regional and state solid waste plans were to take into consideration both present and "reasonably anticipated future needs of the recycling and resource recovery interest within the area encompassed by the planning process."[15] Size was important because the larger the waste-to-energy facility, the more solid waste material it needs in order to operate efficiently. Congress did not want these facilities to be so large that they would monopolize solid waste materials at the expense of other recycling projects. While the tug-of-war over old newspapers will probably continue, it was Congress' hope that the states' solid waste plans could help develop an equitable method of sharing.

Extended Useful Life of Sanitary Landfills

Senator Quentin Burdick (D-North Dakota) was concerned that municipal landfills would run out of space. He offered an amendment to S. 757 directing the EPA Administrator to conduct "detailed, comprehensive studies" on ways to extend the useful life of sanitary landfills and to make better use of sites where filled or closed landfills are located. While the House bill had contained no comparable provision, the Burdick amendment was agreed to by the conferees with the caveat that the agency need not duplicate other studies.

Section 702 of the 1984 Amendments[16] authorizes EPA to conduct demonstration projects, and it directs the agency to "periodically report on the results of such studies." The first of the periodic reports is due to Congress by October 1, 1986.

As part of the Section 702 study, EPA must examine:

- ways of reducing volume of materials *prior* to their being placed in landfills;
- more efficient means of depositing wastes in landfills;

- ways of speeding up decomposition of solid wastes in landfills in a fashion that is "safe and environmentally acceptable";
- methods of producing methane from closed landfill units;
- "innovative uses" of closed landfill sites, such as for energy production (solar or wind) and for metals recovery;
- ways of using sewage treatment sludge to reclaim areas that have been landfilled; and
- coordination of use by one municipality of a nearby municipality's landfill, allowing for establishment of "equitable rates" for that use and considering also the need for providing future landfill capacity to replace the capacity that is consumed.

EPA has indicated that it will not be able to complete this report by the statutory deadline.

Chapter 20 Footnotes

[1] 122 Cong. Rec. S21401 (daily ed., June 30, 1976).
[2] 130 Cong. Rec. S9150 (daily ed., July 25, 1984).
[3] *Id.*
[4] *Id.*
[5] HSWA Sec. 302, adding RCRA §4010(a).
[6] HSWA Sec. 302, adding RCRA §4010(c).
[7] HSWA Sec. 302, adding RCRA §4005(c)(2)(A).
[8] See Conference Report at 80.
[9] *The Environmental Forum*, Sept. 1983, pp.44-48.
[10] 129 Cong. Rec. H8161 (daily ed., Oct, 6, 1983).
[11] *Id.*
[12] *Id.*
[13] Conference Report at 81.
[14] HSWA Sec. 102, adding RCRA §1006(b)(2).
[15] HSWA Sec. 301, amending RCRA §4001; see also HSWA Sec. 301, adding RCRA §4003(d).
[16] HSWA Sec. 702, adding RCRA §8002(r).

Chapter 21

Leaking Underground Storage Tanks: The New Federal Requirements

Summary

Underground Storage Tanks Regulated if:

- Ten percent or more by volume of tank (including connecting pipes) is underground.
- Contains petroleum or hazardous substances listed under Superfund (but not RCRA hazardous wastes which are regulated under Subtitle C).

Exemptions Include:

- Farm or residential tanks with 1,100 gallons capacity or less.
- Tanks used for storing heating oil for consumptive use on premises where stored.
- Septic tanks.

Required Notification of Tank to Designated State or Local Agency:

- Within 30 days of the installation of tanks installed subsequent to November 8, 1985.
- By May 8, 1986, for owner of tanks taken out of operation after January 1, 1974, as well as owners of currently operating tanks.

Other Notification Requirements:

- Beginning 30 days after EPA prescribes the form of notice and for 18 months thereafter, persons who deposit regulated substances in an underground tank must notify owner or operator of such tank of the tank owner's notification requirements.
- Beginning 30 days after EPA issues new tank performance standards, tank vendor must notify purchaser of the tank owner's notification requirements.

EPA Must Issue Regulations:

- By February 8, 1987, for existing and new petroleum tanks.
- By August 8, 1987, for new tanks containing hazardous substances.
- By August 8, 1988, for existing tanks containing hazardous substances.

Regulations Must Cover:

- Maintaining leak detection system.
- Taking corrective actions.
- Reporting leaks and corrective actions.
- Closing tanks to prevent future releases.
- Maintaining evidence of financial responsibility.
- Standards of performance for new tanks.

Other Important Provisions:

- Interim requirements that prohibit installation of tanks that do not meet certain design standards (unless tanks are installed in noncorrosive soil).
- LUST provisions are applicable to tanks owned by federal government.
- Studies required of petroleum tanks by November 8, 1985, of hazardous substance tanks by November 8, 1987, and of exempted farm and heating oil tanks by November 8, 1987.
- States may regulate underground storage tanks if regulations are as stringent as federal requirements.

Congressional Consideration of the Issue

Despite the title, the Hazardous and Solid Waste Amendments of 1984 do more than regulate wastes. Title VI of the statute establishes an entirely new subtitle to the Solid Waste Disposal Act (usually referred to by the name given to the 1976 Amendments, the Resource Conservation and Recovery Act). Title VI of the 1984 Amendments (codified as Subtitle I) mandates that EPA develop a program to regulate underground storage of gasoline and chemical products. Those who have followed the slow progress of environmental legislation in recent years cannot be faulted for expressing some astonishment at the enactment of an entirely new and comprehensive regulatory program in the space of a few short months.

For many decades it has been considered an appropriate safety precaution to store petroleum products and other flammable substances in underground tanks in order to reduce the chances of fire or explosion. However, in the early 1980s numerous reports detailed extensive environmental damages resulting from leaks from underground tanks. It was in response to those reports that Congress enacted the leaking underground storage tank provision (frequently referred to by its acronym, LUST) even though the issue had not been addressed in the original House or Senate RCRA reauthorization measures.

The reports of problems with underground storage tanks that prompted congressional action fell into a familiar pattern in which an undetected leak slowly contaminated the underground drinking water supply for a neighborhood or even entire communities. For example:

> In 1980, a major gasoline leak was traced to a Chevron station in the Northglenn suburb of Denver, Colorado. Chevron was compelled to purchase 41 of the homes in the affluent neighborhood at approximately twice their appraised value. Estimated costs to Chevron are about $10 million.
>
> In Dover-Walpole, Massachusetts, 23 private wells were contaminated with gasoline in 1981. The responsible party, Texaco, has provided the residents with bottled water since then.
>
> At least 3,000 gallons of gasoline leaked from a service station's underground storage tank on the tip of Cape Cod, Massachusetts. Provincetown lost 60 percent of its public water supply when the South Hollow wells in North Truro were shut down in December 1977 to prevent the wells from pulling in contaminated groundwater from the tank site and poisoning the whole well field. More than $1 million had been spent as of the end of 1985 on cleanup and groundwater treatment; approximately $2 million more is needed, and the well field will probably be out of full service for another three to five years.
>
> Leaks of highly toxic solvents from more than 100 underground

> storage tanks in California's "Silicon Valley" have contaminated drinking water in many communities. Thus far, more than $70 million has been spent on cleanup. In October 1984, EPA listed the contaminated aquifers as "Superfund" sites.

Although these examples were dismissed by the American Petroleum Institute as merely "anecdotal" evidence, data compiled by the Congressional Research Service and others indicate that the problem was not isolated. According to the Conference Report, approximately 75,000 underground tanks currently are leaking and up to 350,000 more tanks are likely to begin to leak over the next five years.[1]

Additional evidence of the problem was presented in testimony before Senator Dave Durenberger's Toxic Substances Subcommittee on March 1, 1984, in what turned out to be the only congressional hearing on the subject during the 98th Congress. The subcommittee learned that the principal culprit was corrosion: 66.6 percent of the leaks from piping and 90.7 percent of the steel tanks leaks were caused by corrosion.[2] According to several witnesses, the problem would only get worse since the hundreds of thousands of tanks installed during the 1950s and 1960s (while the demand for gasoline seemed boundless) were now reaching the end of their useful life. Yet most of these tanks still were being filled with gasoline or had been abandoned with gasoline left in them. Because a single gallon of gasoline, according to EPA, could contaminate 50,000 gallons of water, the threat to health could not be ignored.[3]

Armed with these figures (and considerable anecdotal material), advocates of federal legislation pushed for comprehensive regulation. On February 29, 1984, Congressman Don Ritter (R-Pennsylvania) introduced H.R. 4761, a bill to allow Superfund money to clean up leaks from underground storage tanks containing hazardous substances or petroleum products. The bill would also have created an EPA regulatory program for underground tanks. The next day, Senator Durenberger introduced similar legislation as an amendment to the Safe Drinking Water Act.

On the House side, the most obvious legislative vehicle to carry a LUST provision, H.R. 2867 (the RCRA reauthorization bill) already had been passed by the House in November 1983. Consequently on May 10, 1984, when Representatives Florio, Eckart, Dingell, Lent, and Ritter introduced H.R. 5640, a reauthorization of the Superfund program, they included in their bill a free-standing title to establish a comprehensive regulatory program governing underground storage tanks, patterned after H.R. 4761, the original Ritter bill. H.R. 5640 also would have eliminated the "petroleum exclusion" in Superfund, which would have the effect of extending the Act's joint and several and strict liability provisions to those reponsible for leaking gasoline tanks.

On July 25, 1984, after lengthy negotiations with the American Petroleum Institute and other groups representing tank owners, Senator Durenberger offered a less stringent version of his original bill as an amendment to S. 757, the Senate's RCRA reauthorization bill. The Durenberger amendment was adopted by a voice vote, and with the passage of S. 757 the stage was set for a House-Senate conference on the two different RCRA bills. Two weeks later, the House passed its Superfund bill, H.R. 5640, and, in the process, adopted several strengthening amendments to the LUST title offered by Congressman Norman Mineta (D-California).

Since the Senate failed to consider a Superfund bill on the floor, the subject of leaking underground storage tanks could be resolved only in the RCRA conference, where the House conferees could not, at least technically, bargain on the basis of their Superfund LUST provision (which was generally acknowledged to be more stringent than the Senate's). Nevertheless, the House conferees, led by Congressman Eckart, did negotiate on the basis of the Superfund provisions, and the final version of the new RCRA Subtitle represents a clear-cut compromise between the House position in its Superfund bill and the Senate's version in its RCRA bill. It is worth noting that the House provision eliminating the Superfund "petroleum exclusion" was not part of the final LUST compromise.

Scope of Regulatory Coverage: Key Definitions and Exemptions

RCRA Section 9001[4] defines "underground storage tank" as a tank or combination of tanks (including underground pipes which are connected to such tanks) used to contain regulated substances the volume of which is ten percent or more beneath the surface of the ground. Since the volume of the underground pipes must be included in the calculation of the 10 percent threshold, it is entirely possible for the storage tank portion of the system to be located above ground and yet be classified as an underground tank if the connecting underground pipes constitute 10 percent of the overall volume. The term "regulated substance" for purposes of the LUST Title means petroleum, or any petroleum product, or any hazardous substance as defined in Section 101(14) of Superfund,* but *not* a listed or identified hazardous waste under Subtitle C of RCRA. The term "operator" is defined as any person in control of, or having responsibility for, the daily operation of an underground storage tank.

Section 9001 excludes from the definition of underground tanks any:

- farm or residential tank of 1,100 gallons or less capacity used for storing motor fuel for noncommercial purposes;
- tank used for storing heating oil for consumptive use on the premises where it is stored;
- septic tank;
- pipeline facility (including gathering lines) regulated under the Natural Gas Pipeline Safety Act of 1968,[5] the Hazardous Liquid Pipeline Safety Act of 1979,[6] or an intrastate pipeline facility regulated under state laws comparable to the federal statutes;

*Under Section 101(14), hazardous substances include any hazardous substances or toxic pollutants designated under the Clean Water Act, any hazardous air pollutant listed under the Clean Air Act, or any imminently hazardous chemical substance under Section 7 of the Toxic Substances Control Act.

- surface impoundment, pit, pond, or lagoon, storm water or waste water collection system;
- flow-through process tank;
- liquid trap or associated gathering lines directly related to oil or gas production and gathering operations; or
- storage tank situated in an underground area (such as a basement, cellar, mineworking, drift, shaft, or tunnel) if the storage tank is situated upon or above the surface of the floor.[7]

Notification

Not later than November 8, 1985, each owner of an underground storage tank must notify the designated state or local agency of the existence of the tank.[8] The notification, which is to be provided on a form prescribed by the EPA Administrator,* must specify the age, size, type, location, and use of each tank. The same information must be provided (within 30 days of installation) for any tank brought into operation after November 8, 1985.[9]

For each tank taken out of operation after January 1, 1974, the owner must by May 8, 1986, provide notice of the existence of the tank (unless the owner knows the tank subsequently was removed from the ground) and the date the tank was taken out of operation.[10] The notice must also specify the age of the tank; when it was taken out of operation; the size, type and location of the tank; and the type and quantity of any substances left in the tank.[11]

In addition, beginning 30 days after the Administrator prescribes the form of notice (and for 18 months thereafter) anyone who deposits regulated substances in an underground storage tank must notify the owner or operator of the notification requirements.[12] Beginning 30 days after the Administrator issues new tank performance standards, anyone who sells a tank (for use as an underground storage tank) must notify the purchaser of the notification requirements.[13]

Release Detection, Prevention and Correction Regulations

Section 9003[14] requires EPA to promulgate release detection, prevention, and corrective action regulations as may be necessary to protect human health and the environment.** These regulations must include (but need not be limited to) requirements for:

*The final rule concerning notification was published on November 8, 1985. 50 Fed. Reg. 46602.

**Section 207 of the 1984 Amendments, adding RCRA Section 3004(w), directs the Administrator by March 1, 1985, to promulgate regulations under Subtitle C for underground tanks containing hazardous wastes that cannot be entered for inspection. Although EPA proposed regulations for underground hazardous waste tanks on June 26, 1985 (50 Fed. Reg. 26444), the Office of Management and Budget has held up issuance of these standards and the Environmental Defense Fund brought suit to require their issuance. On January 28, 1986, the District of Columbia district court ruled that EPA had failed to perform its mandatory duties under Section 3004(w) to promulgate rules for underground tanks containing hazardous wastes and that OMB has no authority to delay the promulgation of these rules beyond the statutory deadline. The court also ordered EPA to promulgate the regulations by June 20, 1986. *Environmental Defense Fund v. Lee M. Thomas,* Civil Action No. 85-1747 (D.D.C. Jan. 23, 1986).

- maintaining a leak detection system, an inventory control system together with tank testing, or a comparable system or method designed to identify releases in a manner consistent with the protection of human health and the environment;
- maintaining records of any monitoring or leak detection system or inventory control system or tank testing system;
- reporting any releases and corrective action taken in response to a release from an underground storage tank;
- taking corrective action in response to a release from an underground storage tank;
- closure of tanks to prevent future releases of regulated substances into the environment;[15] and
- maintaining evidence of financial responsibility for taking corrective action and compensating third parties for bodily injury and property damage caused by sudden and nonsudden accidental releases arising from operating an underground storage tank.[16]

In promulgating these regulations, EPA is permitted to distinguish between types, classes, and ages of underground storage tanks.[17] In making those distinctions, the agency may take into consideration various factors including but not limited to: location of the tanks, soil and climate conditions, uses of the tanks, history of maintenance, age of the tanks, current industry recommended practices, national consensus codes, hydrogeology, water tables, size of the tanks, quantity of regulated substances periodically deposited in or dispensed from the tank, the technical capability of the owners and operators, and the compatibility of the regulated substance and the materials of which the tank is fabricated.

Deadlines for EPA Tank Regulations

By May 8, 1987, the Administrator must promulgate leak detection, prevention, and corrective action regulations for tanks containing petroleum products.[18] By November 8, 1988, the Administrator must promulgate these regulations for tanks containing hazardous substances.[19]

The Administrator also must establish performance standards for new tanks. By May 8, 1987, those standards must be made effective for new tanks designed to store petroleum products and by November 8, 1987, for new tanks designed to store hazardous substances.[20] The new performance standards must include design, construction, installation, release detection, and compatibility standards, and must be issued no later than three months prior to their effective dates.[21]

Interim Prohibition

Section 9003 also establishes certain "interim" rules, that is, the statutory requirements that will be in effect until EPA promulgates its LUST regulations.[22] Beginning May 7, 1985, no underground storage tank may be brought into use unless the tank will prevent releases resulting from corrosion or structural failure during the operational life of the tank,* and is (a) cathodically protected; (b) constructed of a non-corrosive material (such as fiberglass); (c) clad with a noncorrosive material; or (d) designed to prevent the release of the stored substance. In addition, the material used in the construction or lining of the tank must be compatible with the substance to be stored.

The interim prohibitions do not apply if the tank is installed in soil with resistivity of 12,000 ohms/centimeter or more.** (EPA is authorized to prescribe a more stringent standard.) The soil tests must be conducted in accordance with ASTM Standard G57-78 or another testing method approved by the Administrator.

Inspections, Monitoring and Testing

To help EPA develop and enforce underground tank regulations, Section 9005 mandates that owners and operators provide, upon request, information relating to their tanks and associated equipment, along with all relevant records.[23] In addition, EPA officials (or representatives of a state with an approved program) are authorized to:

- enter at reasonable times any establishment or other place where an underground storage tank is located;
- inspect and obtain samples from any person of any regulated substances contained in such tank; and
- conduct monitoring or testing of the tanks, associated equipment, contents, or surrounding soils, air, surface water or ground water.

Enforcement

Section 9006 authorizes the Administrator to issue orders requiring compliance with any requirement the agency finds is being violated.[24] Failure to comply may result in a civil penalty of not more than $25,000 per tank for each day of continued non-compliance. The Administrator is also authorized to seek injunctive or other appropriate relief.

In addition, any owner or operator who fails to comply with the "interim prohibition" provision or any requirement or standard promulgated by EPA (or a state with an approved program) may be subject to a maximum

*EPA has interpreted "operational life" of a tank to be "the time during which the tank stores regulated substances." EPA Draft Technical Guidance on Interim Prohibition, Aug. 29, 1985, p.I-3. On June 4, 1986, EPA promulgated regulations interpreting various aspects of the Interim Prohibition. For example, the regulations make clear that paint and asphalt coatings are not adequate for cathodic protection. 51 Fed. Reg. 20418, 20420 (June 4, 1986).

**EPA has estimated that of the 100,000 tanks that are replaced each year, 60,000 are for farm and home use and are not subject to RCRA requirements including the interim rules. Of the 40,000 in commercial or industrial use, approximately 13,000 are made of unprotected carbon steel and are subject to the interim rules. However, 75 percent of those tanks "are probably placed in noncorrosive soils" (i.e., where the resistivity is higher than 12,000 ohm/cms). Codification, 50 Fed. Reg. 28741 (July 15, 1985).

civil penalty of $10,000 for each tank for each day of continued violation.[25] The criminal provisions of Section 3008 may not be used against violators. However, according to the Conference Report, the imminent and substantial endangerment provisions (Sections 7002 and 7003) are applicable to leaking tanks.[26]

A key element of Subtitle I is the provision which allows states to administer and enforce the underground storage tank program.[27] Despite the complex procedures involved in authorizing states to manage hazardous waste programs under Subtitle C, Congress duplicated these requirements for the underground storage tank program. Under Section 9004, a state underground storage tank program (which could cover petroleum products or other regulated substances or both) can be approved by EPA if it includes certain basic components and provides for adequate enforcement to assure compliance. Essentially, requirements must be adopted for:

- maintaining a leak detection system, an inventory control system (together with tank testing), or a comparable system;
- maintaining records of a leak detection system or inventory control system;
- reporting releases and taking corrective action;
- closure of tanks to prevent future releases;
- financial responsibility to assure corrective action and compensation of injured third parties; and
- performance standards for new tanks.[28]

Except for financial responsibility requirements, a state program can be approved only if the requirements are "no less stringent" than the corresponding requirements promulgated by EPA under Section 9003.[29] However, EPA is authorized to approve a state program for a limited time even though certain components of the program are deemed to be less stringent.

The amount of time available to a state before it must have comparable provisions depends on whether the provisions must be adopted by the legislature or the state's regulatory agency. If only regulatory action is needed, the state must adopt "no less stringent" standards within one year after EPA has promulgated its regulations. If state legislative action is required, the deadline is two years; if both legislative and regulatory action are needed, the regulations by the state agency must be completed within one year after the legislature has acted. These seemingly complex rules represent a compromise between members of the Conference Committee who wanted to be sure that underground storage tank standards met certain minimum standards, and those who wanted states to proceed with their own programs, rather than wait for EPA to announce the "ultimate" standard.

In addition, Congress allowed states to satisfy the "financial responsibility" component of the program by recognizing "self-insurance pools" or assessing fees on tank owners and operators.[30] Concerned that insurance for "nonsudden" leaks may be unavailable, Congress wanted to specifically provide states with the flexibility to establish "LUST Superfunds" which could be financed by tank owners and operators.

Another important similarity to the Subtitle C program is EPA's affirmative obligation to make sure that an authorized state is doing its job. Thus whenever the Administrator determines, after a public hearing, that a state is not administering or enforcing an approved program, he or she must notify that state.[31] If the state fails to make the necessary improvements within 120 days, the Administrator is required to withdraw approval of the state's program and re-establish the federal program.

Federal Facilities

Like the Subtitle C requirements governing federal hazardous waste facilities, EPA's LUST regulations are applicable to underground storage tanks owned and operated by the federal government. Congressional concern about the federal government's role as a tank owner was based in part on the large number of tanks under its control. The number of operational motor fuel tanks in the continental United States owned by the Department of Defense, for example, has been estimated at more than 40,000.

Congress explicitly provided in the 1984 Amendments that federal agencies would be subject to the same "substantive and procedural" requirements as those imposed on the private sector.[32] The provision also provides that the United States and its officers and employees are not immune "from any process or sanction of any state or federal court with respect to the enforcement of any such injunctive relief."

Unlike the federally owned hazardous waste facilities, however, underground storage tanks owned by the federal government can be exempted from regulation if the President determines that it is in the "paramount interest of the United States" to do so. However, the "paramount interest" exemption cannot be granted for lack of appropriations unless the President specifically requested an appropriation and Congress failed to approve it.

Studies

Since the scope of Subtitle I is not comprehensive (because of the numerous exemptions) and because it was recognized that virtually all underground tanks are susceptible to leaks, Congress directed EPA to conduct a study of all tanks that were not subject to regulation as well as tanks containing petroleum.[33] The petroleum tank study is due by November 8, 1985, and the unregulated tank study by November 8, 1987. Both studies must include an assessment of the ages, types (including methods of manufacture, coatings, protection systems, the compatibility of the construction materials and the installation methods) and locations (including the climate of the locations) of such tanks. The studies must also take into account the soil conditions, water tables, and hydrogeology of tank locations. In addition, the studies must examine the relationship between these factors and the likelihood of releases from underground storage tanks. Finally, the studies are to review the effectiveness and costs of inventory systems, tank testing, and leak detection systems, "and such other factors as the Administrator deems appropriate."

The same deadline of November 8, 1987, applies to an additional study regarding farm and heating oil tanks. This study must include estimates of the number and location of such tanks and an analysis of the extent to which there may be releases or threatened releases from these types of tanks.

Although Subtitle I of RCRA mandates an ambitious and far-reaching regulatory system, EPA and the states appear to be committed to carrying out the letter and spirit of the new law. This task will be aided by the recognition that a leaking underground tank is not in anyone's interest. Although prudent self-interest was not sufficient to prevent the problem, it is likely to be a strong element in its solution, particularly since the potential liability for an underground leak is enormous. Over the next four years, tank owners, tank manufacturers, leak detection specialists, and other experts have an opportunity to share their knowledge with EPA and states and help develop a workable program. Guided by the common purpose of preventing leaks and by the explicit statutory mandate to protect human health and the environment, there appears to be no reason why American expertise and ingenuity cannot solve the problem of 350,000 leaking underground storage tanks.

1986 Superfund Amendments

Several major revisions to Subtitle I were made by the Superfund Amendments and Reauthorization Act of 1986 (Public law 99-499). In enacting these changes, Congress' foremost objective was to provide a mechanism for responding to leaks from underground petroleum tanks without repealing Superfund's "petroleum exclusion." The solution Congress devised was to create a $500 million "mini-Superfund," financed by a one-tenth of a cent tax on gasoline and other motor fuels, which would be available to pay for clean-up costs whenever the leaking tank's owner or operator either could not pay or could not be located. However, in order to avoid depleting the Leaking Underground Storage Tank Fund, Subtitle I was amended to impose strict and joint and several liability on tank owners and operators in cost recovery actions and to require the Administrator to establish *mandatory* financial responsibility requirements. Enforcement of the financial responsibility obligations may be waived temporarily for categories of tanks (rather than individual owners) if it is demonstrated that insurance is unavailable and progress is being made to establish a risk retention group or a state-funded insurance pool. In addition, the Administrator is given explicit authority to order tank owners and operators to undertake corrective action. It is anticipated that states will assume the primary task of implementing the clean-up and enforcement components of the revised underground tank program.

Chapter 21 Footnotes

[1] Conference Report at 128. See also testimony by Jack E. Ravan, EPA Assistant Administrator, hearings before the Senate Subcommittee on Toxic Substances and Environmental Oversight, Rpt. No. 721, 98th Cong., 2d Sess., 73. Subsequent studies indicate the problem of leaking tanks may be much bigger than originally anticipated. For example, EPA's Office of Toxic Substances conducted a study which concluded that up to one-third of the nation's 750,000 underground tanks containing more fuel may be leaking. See *"Hazardous Waste Report,"* June 9, 1986, at 5.

[2] Senate hearings, supra note 1, 276.

[3] *Id.* at 73 and 276.

[4] HSWA Sec. 601(a).

[5] 49 U.S.C. §§1671, *et seq.*

[6] 49 U.S.C. §§2001, *et seq.*

[7] HSWA Sec. 601(a), adding RCRA §9001(1).

[8] HSWA Sec. 601(a), adding RCRA §9002(a)(1).

[9] HSWA Sec. 601(a), adding RCRA §9002(a)(3).

[10] HSWA Sec. 601(a), adding RCRA §9002(a)(2)(A).

[11] HSWA Sec. 601(a), adding RCRA §9002(a)(2)(B).

[12] HSWA Sec. 601(a), adding RCRA §9002(a)(2)(B)(5).

[13] HSWA Sec. 601(a), adding RCRA §9002(a)(2)(B)(6).

[14] HSWA Sec. 601(a).

[15] HSWA Sec. 601(a), adding RCRA §9003(c).

[16] HSWA Sec. 601(a), adding RCRA §9003(d).

[17] HSWA Sec. 601(a), adding RCRA §9003(b).

[18] HSWA Sec. 601(a), adding RCRA §9003(f)(1).

[19] HSWA Sec. 601(a), adding RCRA §9003(f)(3).

[20] HSWA Sec. 601(a), adding RCRA §9003(f)(2).

[21] HSWA Sec. 601(a), adding RCRA §9003(e).

[22] HSWA Sec. 601(a), adding RCRA §9003(g).

[23] HSWA Sec. 601(a).

[24] HSWA Sec. 601(a), adding RCRA §9006(a).

[25] HSWA Sec. 601(a), adding RCRA §9006(d)(2)(C).

[26] Conference Report at 127.

[27] HSWA Sec. 601(a), adding RCRA §9004.

[28] HSWA Sec. 601(a), adding RCRA §9004(a)(2).

[29] HSWA Sec. 601(a), adding RCRA §9004(b).

[30] HSWA Sec. 601(a), adding RCRA §9004(c).

[31] HSWA Sec. 601(a), adding RCRA §9004(e).

[32] HSWA Sec. 601(a), adding RCRA §9007.

[33] HSWA Sec. 601(a), adding RCRA §9009(a) and (b).

Chapter 22

The Fight Over Law Enforcement Authority for EPA

The final chapter on the 1984 Amendments is not intended as an analysis of one of the Amendments' provisions. Rather, it describes one of the few floor fights that took place, thereby conveying some of the flavor of the legislative process leading to the enactment of the 1984 Hazardous and Solid Waste Amendments.

When a controversial bill is debated on the floor of the House or Senate, it is relatively rare for the participants to engage in an all-out, unrelenting effort to win. Usually this type of combat can be avoided through a last minute compromise. Sometimes it's just not worth the trouble, because the outcome of the vote is known well in advance. On those occasions, the debate is less heated, the hyperbole less inflated, and the syntax less mangled than in a real floor fight.

In a real fight—one in which the result cannot be predicted, with bands of fiesty and determined legislators on either side—the fireworks are worth watching. During the reauthorization of the Resource Conservation and Recovery Act, there were a number of *real* floor fights—all on the House side. In retrospect, the issues provoking these fights were of far less import than the major components of H.R. 2867. Nevertheless the House became embroiled in such questions as:

- Should generators who produce less than 100 kilograms of waste a month (but more than 25 kilograms) be required to notify the transporter that the waste is hazardous?
- If the Department of Justice refuses or declines to prosecute a case referred by EPA, should EPA be allowed to prosecute it?
- Should Congress approve EPA's small quantity generator regulations before they can become effective, giving Congress, in effect, a veto over Executive Branch rulemaking?

The floor fight generating some of the loudest rhetorical fireworks was the one over guns—specifically whether EPA's criminal investigators should be authorized to carry arms when investigating midnight dumping and other crimes involving hazardous waste.

As passed by the Energy and Commerce Committee, H.R. 2867 contained a provision entitled "Law Enforcement Authority." That provision, among other things, would have given EPA criminal investigators the right to carry firearms, make arrests, and execute and serve subpoenas and warrants. This authority has been conferred by Congress not only on FBI agents, but also on law enforcement personnel in other federal agencies such as the Drug Enforcement Administration and the Fish and Wildlife Service. Given the widespread concern over midnight dumping of hazardous waste, the inherent dangers involved in policing "midnight dumping," and the fact that other agencies possessed law enforcement authority, it seemed appropriate for EPA's criminal investigators to have the same powers.

But the House Judiciary Committee saw things differently. That Committee had been trying for several years to end the "proliferation" of this type of authority, and it finally decided it was going to draw the line at EPA.

On June 17, 1983, one month after the Energy and Commerce Committee approved H.R. 2867, the Judiciary Committee voted to delete the provision on law enforcement. With two major committees in direct conflict, the stage was set for a major legislative battle on the House floor—a fight that nobody, including the leadership of the House, made any effort to avoid. On November 3, 1983, the fourth and final day that the House debated the RCRA Reauthorization bill, Representative William Hughes (D-New Jersey), the chairman of the Subcommittee on Crime, offered an amendment on behalf of the Judiciary Committee to strike the offending provision. In his opening remarks, Hughes quickly invoked the support of the Reagan Administration and specifically of the Department of Justice, stating:

> The Department feels strongly that the granting of these authorities should be strictly limited in order to protect both citizens and law enforcement officials. The Department feels that it is important that law enforcement officers have specific training and that the more law enforcement authority is expanded and given to other agencies, the more likelihood there is for problems in the use of that authority.[1]

Hughes also raised the spectre of coordination problems, stating:

> The Department will be less able to coordinate law enforcement matters as EPA assumes these tasks independently. It is a matter of high priority in the Department to investigate waste removal cases. EPA's independence may result in the Department and EPA working on the same case without either agency being aware of the actions of the other. Coordination between the two agencies, each doing what it is trained to do, is the most effective law enforcement scheme.[2]

After Hughes outlined the case for the Judiciary Committee's amendment, the chairman of the Committee, New Jersey Democrat Peter Rodino, told the House that the FBI had made toxic waste cases a "special priority" and that EPA and the Justice Department "have an excellent working relationship."[3] Hughes then distilled the Judiciary Committee's argument into two sentences:

> You cannot have a police department, a minipolice department, in every agency and department of Government and expect to have some centralized, coordinated attack upon the crime problems that exist in every aspect of these departments and agencies. It is as simple as that.[4]

The third participant in the debate, the principal author of H.R. 2867 and also from New Jersey, was Representative James J. Florio. He vigorously defended the law enforcement provision, claiming that because provisions of this bill restrict land disposal of hazardous waste and require safer methods of treatment and recycling, there will be a significantly greater incentive to dispose of toxic waste illegally.[5] "By voting for this amendment," Florio continued, "we will be depriving EPA of the enforcement tools necessary to protect against deliberate criminal violations. That is not a signal that this Congress should send to the American people. I strongly urge my colleagues to defeat this amendment."[6]

Florio then entered into the record a list summarizing the law enforcement credentials of EPA's 22 criminal investigators. The list was designed to show that "we are talking about professionals in the field of criminal justice."[7] The list, which had been prepared by EPA staff, showed that each of EPA's investigators had a minimum of five years of criminal enforcement experience, and most had at least 10 years' experience.

Representative Harold Sawyer (R-Michigan), the senior Republican member of the Judiciary committee, wanted to draw the line on enforcement authority "proliferation." He responded that the time had come to say no to an agency:

> [E]veryone wants their own mini police department, all running around carrying guns, making warrantless arrests. . . . There is not an agency in the Federal Government that does not have some kind of enforcement problem. . . . The Federal Communications Commission, they are dealing with radios or radio stations, private or otherwise, there [sic] are not licensed. I suppose they might want to have their own armed policemen riding the circuit, with radio direction finders, making arrests; the Federal Trade Commission, certainly in the antitrust field, out with guns and weaponry; the Interstate Commerce Commission, on motorcycles, rounding up and arresting illegal truckers operating on the road without ICC certificates of convenience and necessity.[8]

Fellow Republican Guy Molinari of New York disagreed, pointing out that FBI agents could not always be around when they might be needed. Molinari felt that, if EPA's criminal investigators were pursuing a case and "they come upon a dumping activity, they should be empowered to make an arrest; they should be empowered to protect themselves . . . [I]f we are going to declare war on those who are abusing the environment . . . then we should be ready to protect those who protect us." [9]

Florio was ready with an example to underscore Molinari's point. He cited a report from the Seattle office of EPA's Criminal Enforcement Division.

> EPA investigator witnessed a truck leaking PCB liquids on the highway between Portland and Seattle. Investigator unable to stop truck or prevent contamination of 50 miles of highway before the truck was ultimately stopped by a sheriff who was able to be obtained for purposes of enforcing the law.[10]

Representative Norman F. Lent (R-New York) recounted an incident in Portland, Oregon, in which the target of a RCRA investigation had "threatened to kill business associates and dispose of [their] remains in hazardous waste drums."[11]

Representative Ron Wyden (D-Oregon) tried to undercut the Administration's endorsement of the Judiciary Committee amendment. Wyden quoted from an April 1983 letter from EPA's Associate Enforcement Counsel to the Justice Department requesting that EPA's investigators be "deputized" as Assistant U.S. Marshals—a designation which would give them law enforcement authority:

> Without full law enforcement authority, significant limits will be placed on EPA's criminal initiative. More importantly, the program will not be able fully to ensure the safety of its investigative staff, and of citizens that choose to assist EPA in pursuing allegations of criminal misconduct. . . . It is my personal belief that law enforcement powers should be pursued for EPA's criminal investigators immediately.[12]

Hughes, who at that point was not getting much support from his colleagues, reiterated his concern about proliferation of law enforcement authority: "We cannot possibly have a centralized system if we have a mini-police department in every agency and department. That is what we are moving toward."[13]

Representative Dennis Eckart (D-Ohio) rose in opposition to the amendment and emphasized the nature of the threat to EPA's investigators:

> It is widely known that we are not dealing with a bunch of panty-waisted amateurs who view this midnight dumping as their recreational activity. . . . What we are dealing with is a highly profitable enterprise that has been infiltrated at the highest levels by organized crime and which exposes those whose jobs are to protect both the environment and our citizenry from criminal activities to very egregious and serious threats. . . . In Pennsylvania, a landfill operator shoots two policemen, one fatally. In Chicago, a landfill operator runs his bulldozer up against an inspector. In Rockland County, the sheriff's office, a lieutenant is shot at. In Kentucky and Georgia, inspectors are forced away from the site at gunpoint. In Ohio, one of our inspectors is told: "Guys with jobs like yours can get shot."[14]

Representative Tom Kindness (R-Ohio) said the debate was a "repetition of the debates that have occurred before . . . where the same arguments . . . the same horror stories are presented every time." According to Kindness:

> The people change, the faces change, the names change, the particular incidents, but it is the same sort of thing: This particular case is so important that we must set up a separate police force for it, a separate, uncoordinated police force.[15]

In response to Kindness, Representative John Dingell (D-Michigan), the influential Chairman of the Energy and Commerce Committee, pointed out that EPA had requested help from the Department of Justice and had been ignored. Trying to turn the "proliferation" argument in favor of extending law enforcement authority to EPA, Dingell emphasized that EPA was not "asking for special authority. They are asking for the same authority that tick inspectors for the Department of Agriculture have to carry out the law."[16]

Rodino, re-entering the debate, said he agreed there should be law enforcement but insisted that "this law enforcement ought to be predicated on need, on the need to deputize these individuals. . ."

Dingell wanted to know, in that case, why the request for deputizing EPA's investigators had not been acted on for nearly six months—a question Rodino could not answer. At this point, the distinguished and outspoken chairman of the Energy and Commerce Committee became visibly angry, noting:

> The Department of Justice has done nothing, nothing, and we are not talking about people that play patty cake. We are not talking about ladies' bridge clubs. We are not talking about seminarians, but people that shoot big black guns and firebomb other folks' businesses.[17]

After several minutes of debate in which the same arguments were rehashed, Hughes, perhaps fearing the imminent defeat of the Judiciary Committee amendment, offered a "perfecting" amendment which the clerk read as follows:

> Perfecting amendment to the Judiciary Committee amendment offered by Mr. Hughes: Page 33, strike out 1 and all that follows down through line 12 on page 34 and substitute:
>
> (e) LAW ENFORCEMENT AUTHORITY.—(3) The Attorney General shall, at the request of the Administrator (and on the basis of a showing of need) deputize qualified employees of the Environmental Protection Agency to serve as Special Deputy United States Marshals in criminal investigations with respect to violations of the criminal provisions of the Act.

The amendment, which Hughes and the Judiciary Committee staff had drafted only minutes before, appeared to be a concession to the Energy and Commerce Committee. However, as a practical matter, the perfecting amendment did nothing to alter the *status quo*. The Attorney General would not be forced to deputize anyone and no one was more aware of this reality than John D. Dingell. Questioning the author of the perfecting amendment, Dingell asked, "Can the gentlemen tell me what the difference between the amendment he offers now and existing law is?"

> Mr. HUGHES. Well, the difference is that this is now mandatory, this says the Attorney General shall.
>
> Mr. DINGELL. It is not, it is mandatory upon showing of need.

> Mr. HUGHES. Upon the showing of need.
>
> Mr. DINGELL. What is a showing of need?
>
> Mr. HUGHES. I would say to my colleague that if in fact the Administrator of the EPA has communicated to the Attorney General that there is a specific investigation under way which requires a special deputization and substantiates that with the nature of the continuing investigation, presenting evidence of a risk to the safety of the individuals pursuing the investigation or showing that in fact the groups have some connection to organized crime as has been raised during this debate, then under those circumstances the Attorney General must deputize.

Shortly after that last skirmish, the House voted. Most observers felt that the Energy and Commerce Committee had dominated the two hours of debate. There was no doubt that the committee members on the floor— Dingell, Florio, Eckart, Gore, Lent, and Wyden—had taken full advantage of the emotional issue of hazardous waste disposal, linking it, wherever possible, to the equally volatile issue of organized crime. But when the vote was counted it was not even close: Hughes' perfecting amendment was adopted 292 to 125. It was an overwhelming victory for the Judiciary Committee and a stunning defeat for the Committee on Energy and Commerce.

What went wrong? Although some members of the House were persuaded that law enforcement authority should not be further "decentralized," this argument was probably not as compelling as the Judiciary Committee thought it was. The more likely explanation involves a subjective evaluation of John Dingell: many members of Congress felt that Chairman Dingell, an active member of the National Rifle Association, was not just trying to extend law enforcement authority to EPA; he was attempting to put guns in the hands of every federal employee with a law enforcement function. However, for many Members, the key question was the competence of EPA, which was not, at that point in time, held in high esteem. As one congressman said privately, "I'm not about to give Rita Lavelle a six-shooter."

Several months after the debate, but prior to the enactment of the RCRA reauthorization, Attorney General William French Smith deputized half of EPA's investigators as deputy U.S. Marshals. Currently there are 35 criminal investigators at EPA—all of them have been deputized.

Chapter 22 Footnotes

[1] 129 Cong. Rec. H9133 (daily ed., Nov. 3, 1983).
[2] *Id.*
[3] 129 Cong. Rec. H9134 (daily ed., Nov. 3, 1983).
[4] *Id.*
[5] *Id.*
[6] *Id.*
[7] *Id.*
[8] 129 Cong. Rec. H9135 (daily ed., Nov. 3, 1983).
[9] *Id.*
[10] *Id.*
[11] 129 Cong. Rec. H9136 (daily ed., Nov. 3, 1983).
[12] *Id.*
[13] 129 Cong. Rec. H9137 (daily ed., Nov. 3, 1983).
[14] 129 Cong. Rec. H9138 (daily ed., Nov. 3, 1983).
[15] 129 Cong. Rec. H9139 (daily ed., Nov. 3, 1983).
[16] 129 Cong. Rec. H9141 (daily ed., Nov. 3, 1983).
[17] 129 Cong. Rec. H9142 (daily ed., Nov. 3, 1983).

Index

About the Authors

CHRISTOPHER HARRIS practices environmental law with the firm of Schmeltzer, Aptaker & Sheppard in Washington, D.C. From 1979 to 1983 he worked in the Land and Natural Resources Division of the U.S. Department of Justice, principally on Superfund and Resource Conservation and Recovery Act liability issues. During the 98th Congress he served as Counsel to the House Energy and Commerce Subcommittee which launched the 1984 RCRA Amendments.

WILLIAM L. WANT practices environmental law in Charleston, South Carolina. He is Chairman of the South Carolina Bar Hazardous Waste Subcommittee and a member of a state-wide commission chaired by the Lieutenant Governor on how hazardous waste should be managed in the state. From 1973 until 1982 he was a trial attorney for the Land and Natural Resources Division of the U.S. Justice Department. He has litigated for the Sierra Club Legal Defense Fund and the National Wildlife Federation, taught environmental law at the University of Maryland and George Washington University law schools, and served as General Counsel of the Hazardous Waste Treatment Council. He is the author of numerous articles on environmental law.

MORRIS A. WARD is Director of Environmental and Occupational Health for the American Electronics Association. Previously, he was the founder and editor of *The Environmental Forum.* From 1979 to 1981 he served as Assistant Director of the National Commission on Air Quality. A frequent lecturer on issues relating to environmental protection, he was a news analyst and commentator for National Public Radio's "All Things Considered" and "Morning Edition" programs from 1978 to 1984.